Coal and Crisis

Coal and Crisis

The Political Dilemmas of Energy Management

Walter A. Rosenbaum

PRAEGER PUBLISHERS
Praeger Special Studies

New York   •   London   •   Sydney   •   Toronto

Library of Congress Cataloging in Publication Data

Rosenbaum, Walter A
    Coal and crisis.

    Includes bibliographical references and index.

    1. Coal trade--United States. 2. Coal trade--
Environmental aspects--United States. 3. Energy policy--
United States. I. Title.

HD9546.R67   1978   338.2'7'20973   78-8606
ISBN 0-03-042596-4

PRAEGER SPECIAL STUDIES
383 Madison Avenue, New York, N.Y. 10017, U.S.A.

Published in the United States of America in 1978
by Praeger Publishers
A Division of Holt, Rinehart and Winston, CBS, Inc.

89   038   987654321

Contents

LIST OF TABLES

LIST OF ACRONYMS

BTU	British thermal unit
CED	Committee on Economic Development
CBO	Congressional Budget Office
CEQ	Council on Environmental Quality
ERDA	Energy Research and Development Administration
EPA	Environmental Protection Agency
FEA	Federal Energy Administration
FPC	Federal Power Commission
GAO	General Accounting Office
NCA	National Coal Association
NEP	National Energy Plan
NMC	National Minerals Congress
NGPRP	National Great Plains Resources Program
NRC	Nuclear Regulatory Commission
OTA	Office of Technology Assessment
PCEA	President's Council of Economic Advisors
R&D	Research and development
UMW	United Mine Workers

Coal and Crisis

1

The Ambiguous Abundance: Political Elements in Policy Formation

In America's energy crisis there is both a requiem and a renaissance. For the U.S. frontier, it is a requiem, the irrefutable evidence that the abundance of resources once symbolized by the American west is rapidly ebbing. For the U.S. coal industry, once apparently destined to a twilight of increasing technological obsolescence, the crisis seems to promise rejuvenation. Coal, the only energy inheritance not substantially depleted by the nation's hundred-year raid on resources, has become its most important short-term energy capital. Within the next few years, public institutions and officials must decide how this inheritance will now be spent.

To burn enormous new quantities of coal will be a costly incandescence— vastly so in environmental hazards, capital investment, technological commitments, and further resource utilization. But hoarding coal seems to invite unacceptable risks economically, technologically, and internationally. Thus, the federal government, like most energy planners, favors more and more a massive increase in national coal production. Today, a major issue facing the nation is whether its public institutions, in whose hands the management of any future coal utilization policy will largely rest, can discharge this responsibility with prudent regard for both the benefits implicit in new coal production and the potential dangers. More than a matter of judicious choice among the major policy options is involved. An equally fundamental matter is that public institutions and their rules for making decisions must also be substantially modified; the political formula for resource allocation will be as much an issue in future coal policy as the substantive options. American public institutions have historically allocated resources according to rules reflecting the presumption of continuing resource abundance. This frontier mentality powerfully shaped the American political imagination and contributed to the logic of public policymaking at all levels of American public life. Since this mentality nurtures a political style increasingly inappropriate to current energy management, a wise coal policy requires the passing of the frontier that exists in American political consciousness, just as the actual American frontier west itself has vanished.

The interaction between the nation's political institutions, its political values, its historic experience, and the substantive questions of coal management will be a recurrent theme. This political context is so essential to illuminating the issues associated with national coal utilization that it will be helpful to examine briefly those political elements figuring most conspicuously in future coal policy formulation.

THE POLITICAL CONTEXT: AN OVERVIEW

Until recently, coal was the least politicized of the nation's energy sectors. Aside from regulating mine safety, labor relations, and private access to coal on federal lands, Washington largely left coal in the hands of the private sector.[1] The sudden emergence of coal as a strategic national energy resource will draw the nation's political structure much more directly into future coal management. Several elements in the political system are likely to assume particular salience in shaping the nation's future coal utilization although, as we shall observe, they seldom operate independently in policy formation:

the substantive policy issues which have been successfully promoted to the public agenda. In making public policy, there is a politics of agenda setting. It is a competitive, pluralistic struggle among publics to promote their interests to the policy agenda of public decision-makers; there are winners and losers. Officials seldom have sufficient time or resources to ponder the full implications of a policy question; critical issues, when not skillfully advanced, can be ignored. Given the selectivity inherent in public policy debate, a major factor in future coal policy will be which groups can mobilize sufficient political resources to gain representation on the policy agenda—in effect, who wins the struggle to define the essentials of coal policy "realities" for public officials.

the existing political formula. New coal policy will not emerge from a historical vacuum. The nation already has a political formula for coal utilization. It includes existing governmental agencies (primarily federal) with jurisdiction over coal utilization, the current statutory responsibilities these agencies discharge in coal management, the constellation of organized interests active in coal affairs, and understandings—informal but influential—shared among political actors concerning how coal problems ought to be handled. Equally importantly, Americans share cultural values, deeply rooted but often unarticulated, about the proper treatment of resources. All these elements, the accretion of national experience with coal, are active forces in shaping new coal policy.

strategies of political management. The means through which coal policy goals and priorities are implemented become influential themselves in shaping that policy. Generally, governmental management of future coal will be accomplished through some combination of three approaches: regulatory procedures, technology development, and distributive formulas. Each of these will be examined in subsequent discussion. An important consideration generally is that any technique involves social costs, benefits, and risks. Thus, it is important when assessing policy decisions to illuminate how specific implementation strategies become factors in policy outcomes.

the opinion climate. Public attitudes about energy, and coal in particular, create an opinion climate to which public officials are responsive. The public's knowledge about energy problems, its alertness to options and official responses, and the distribution of such opinions socially and geographically will create a public mood congenial to some policy options and hostile to others. Public opinion seldom dictates policy directly, yet its persistent influence in official deliberations is sufficiently strong that it must be considered a major factor in policy outcomes.

An inventory of politically significant factors so broad in scope and abbreviated in description can only suggest the complexity of coal policy formulation. While the major policy options facing public officials will be the primary consideration, this survey does illuminate a fundamental aspect of the coal debate: the nature and consequences of policy choices cannot be intelligently apprehended without sensitivity to the cultural values, institutional arrangements, and implementation strategies which attend such choices.

The beginnings of the current coal debate lie in the precarious status of U.S. energy reserves and, especially, the enormous presence of coal in all assessments of future energy options. This, the physical context in which all policy discussion occurs, is properly the place to start in understanding why coal no longer lingers at the periphery of the governmental policy agenda.

COAL: PHYSICALLY ABUNDANT, INTENSELY CONTROVERSIAL

The energy crisis is a reality; to believe otherwise is to invite a national energy catastrophe in less than a generation. Estimates as to when this would occur vary, sometimes greatly, due to incomplete data and differing formulas. Nonetheless, the major conclusion of virtually all public and private agencies examining the current U.S. energy economy is that U.S. domestic energy consumption continued at present levels will surge so far beyond domestic petroleum and natural gas reserves that a national emergency will be imminent in a few years. Significantly, studies by the Congressional Budget Office (CBO), the General Accounting Office (GAO), and the Office of Technology Assessment (OTA), otherwise critical of President Carter's National Energy Plan (NEP) in many respects, accept this crisis as a first principle.[2] The OTA summarized the consensus:

TABLE 1.1

Domestic Fossil Fuel Reserves and Their Use, 1977

Fuel	Percent of U.S. Reserves	Percent of World Reserves	Percent of U.S. Energy Use
Coal	90	33	18
Natural gas	4	10	30
Petroleum	3.7	5.2	27

Sources: U.S. Executive Office of the President, Energy Policy and Planning, *The National Energy Plan* (Washington, D.C.: Government Printing Office, April 1977); *New York Times*, July 3, 1977.

> ... the United States and all other industrial nations of the world face a serious energy problem. If the United States tries to escape short-term sacrifices that can begin to deal with the problem, it will face real hardships no more than 10 years from now. ... the basic U.S. problem is a case of domestic demand outstripping domestic supply.[3]

In April 1977 President Carter summoned the nation to an energy account-ability it neither welcomed nor understood. Like most energy planners, the president saw the "near-term" solution—that is, the solution through 1983—in balancing energy demand and supply, not primarily through conservation (though gestures were made in that direction), but in substituting coal for much existing oil and natural gas consumption. The president's call for new coal production, summarized in a single paragraph, became the rallying signal of all interests pressing for coal industry expansion:

> *Resources in plentiful supply should be used more widely as part of a process of moderating use of those in short supply.* Although coal comprises 90 percent of the United States total fossil fuel reserves, the United States meets only 18 percent of its energy needs from coal. ... This imbalance ... should be corrected by shifting indus-trial and utility consumption from oil and gas to coal.[4]

This production call fell upon an industry then hovering near stagnation. Once it had been "King Coal." As late as 1930, coal accounted for 60 percent of U.S. fuel consumption. But production steadily declined until 1960 and thereafter fluctuated around 20 percent of domestic energy consumption as petroleum and natural gas, with their greater fuel efficiency and economic and logistical attrac-tions, drove coal from numerous markets. Coal remains, nonetheless, the only abundant domestic energy reserve. Its multiple attractions to energy planners pre-occupied with an impending energy scarcity can be appreciated by the summary statistics in Table 1.1, which depict coal's status in comparison to other domestic and international fossil fuels. The United States possesses about one-third of the world's coal reserves; except for the Soviet Union, only the United States can contemplate the substitution of coal for petroleum or natural gas and thus can anticipate a potential reduction in currently projected gas and oil consumption. Estimates of this future saving vary, but generally suggest that compulsory sub-stitution could save between 2.4 and 3 million barrels of oil per day, possibly one-third of the anticipated 1985 oil imports.[5]

Many energy experts assert that coal should be considered a "bridge" fuel, used intensively only during a period between the present oil-dependent Ameri-can economy and a future era, generally after the year 2,000, when nonfossil fuels will be major energy suppliers. In this perspective, the nation would be making its coal commitments only as an expedient and temporary strategy while it develops the more desirable nonfossil fuels. This very benign interpretation of coal utilization is customarily advanced to blunt the force of warnings against coal's hazards.

The military attractions of coal are equally important. A powerful stimu-lant to current governmental apprehension over domestic energy consumption is the recognition that, if present energy trends were to continue, by 1985 the United States would have to import at least 52 percent of its daily oil consump-tion. This long-term dependence, notes the CBO, "creates national security risks and makes our economy vulnerable to shocks from outside, especially because the supply and price of oil are to a great extent dictated by an international cartel."[6] This bleak prospect, with all its implied diplomatic and military disad-vantages, appears unacceptable to national leaders. Coal would appear to diminish this dependence; it would also guarantee an important energy source wholly within the continental United States and therefore invulnerable to blockade. In this respect, as U.S. military and diplomatic strategists are quick to argue, coal production assumes immediate security implications. "In the case of the United States, and probably it alone of all the leading industrial states of the free world, diminished supply vulnerability could result from increased coal production."[7]

Coal utilization also has its technological attractions. Coal production tech-nology is relatively simple, proven, and largely in place at numerous existing sites; although rising production would require an extensive new geographic dispersion of coal production facilities, there is no significant lead time necessary to develop the production technology as there is with new domestic oil or gas reserves, nuclear energy, and most other energy sources considered for the near term. Nor does capitalization of production facilities—a problem that seriously bedevils future nuclear energy facilities—appear troublesome because the nation's major petroleum companies, which own a substantial portion of the coal industry, appear able to provide sufficient capital for expanding plants. In fact, controversy over the feasibility of producing coal sufficient to diminish significantly future oil and gas consumption concerns less the capacity to produce enough coal than the ability to transport it to markets at desirable volume and price.

Balanced against these attractions are numerous actual and potential risks. The most widely and intensely voiced objection to extensive new coal production is that pervasive environmental damage may result because regulatory procedures to moderate environmental impacts may not, and perhaps cannot, be applied. There is little doubt that mounting coal production and utilization across American society would produce ominously increased environmental degradation at both production and consumption sites unless rigorous controls were initiated. Strip mining, the principal source of present and future coal, is itself environmentally malevolent without regulatory constraints; additionally, coal burning in industry and utilities produces a variety of dangerous air pollutants. Controversy also surrounds the technical feasibility and economic acceptability of pollution control devices for coal-burning units. Matters of social equity also arise out of real, or potential, air and water contamination: which states and localities should bear the costs, and the inherent risks, of future coal production? Equally significant, many experts assert that increased coal production may ultimately force a tradeoff with existing, or future, air and water quality standards. Pessimists assert that present quality standards may be incompatible with more coal utilization. A more restrained assessment is that further coal utilization will, in effect if not by intent, set the limits of future environmental quality goals. The CBO argues:

> The stringent application of environmental regulations, coupled with the desire to increase dramatically the use of coal in all energy sectors, creates a paradox. Furthermore, should environmental goals become even more ambitious, it is possible that expanded use of coal will become relatively less desirable.[8]

In short, the nation may not be able to enjoy more coal *and* higher air or water quality standards, for one option might constrain the other.

A second broad controversy swirls less around the immediate environmental impact of coal production than about the long-range planning implications. Essentially, many observers assert that a commitment to massive new coal production will spawn a proliferating variety of hard energy technologies adapted to coal utilization and to fuels synthesized from coal but that they are far less desirable than soft technologies still experimental or limited in production. The soft technologies are assertedly more desirable environmentally, economically, and scientifically. This hard/soft debate, as it is known, deals as much with the bias of the political and economic sectors as with energy technologies.[9] Proponents of soft technologies believe that public officials, energy lobbies, and major business interests favor long-term development of hard technologies because of their greater capitalization, public investment, and stimulation to the production sector. Even if a conscious decision is not made to develop long-term hard technologies, assert the critics, such development may be unavoidable once apparently short-term commitments to coal production are made. Thus, warn proponents of restrained development, coal production in the hand of public agencies may become not a bridge fuel to a new, more restrained future energy economy but, instead, a highway leading away from energy conservation and toward increasing energy generation stimulated by hard technology systems.

A third major coal issue relates to the political feasibility of regulatory programs intended to mitigate the environmental damage of coal production and utilization. The passage of the Strip Mining Control and Reclamation Act of 1977 created, for the first time, a national regulatory program for strip-mined coal, largely administered through the states within a framework of standards ordained by federal law. Long overdue and admirable in intent, the new regulatory program is nonetheless susceptible to many administrative ills which reduced most state mine regulation programs to a charade. Moreover, imposing a two-tier regulatory structure on strip mines in which federal agencies exercise vigilance over state agencies which, in turn, implement strip-mine controls may compound enforcement problems. Further, some experts believe that the new bill suffers less from deficiencies in detail than from a grave error in conception. According to these observers, efforts of the federal government to regulate in detail the production processes of major industries are ordinarily ineffective, or achieve grossly suboptimal results, because they depend too much on administrative intervention and rule-making in areas where government is not adequate to the task.[10] Citing the inflation of federal regulatory programs from a handful in the 1950s to more than 70 in the late 1970s, these critics assert that regulatory bureaucracies have themselves become a major problem best solved by other procedures, including reliance on market mechanisms to control sectors of the economy affecting the public interest.

Finally, the matter of national economic growth rates permeates all issues of coal management. It is impossible for energy planners to formulate a future coal policy without making assumptions about the desirable rate of national economic growth. Indeed, the substance of coal policy—for instance, how much new coal production is acceptable in ten years—will itself influence such growth rates. Comprehensive national energy planning, of which coal management is a portion, has given government an opportunity to influence forcefully and deliberately future growth patterns. Advocates of moderate future economic growth, or controlled growth, often assert that national coal policy, like other energy strategies, should include explicit commitments for moderating future economic growth and that future energy consumption plans should be sensitive to these commitments. These observers assert, additionally, that future coal planning should be biased toward resource conservation; this would also inhibit large increments in the nation's economic growth rate.

The major controversies which collectively constitute the coal issue all, directly or implicitly, concern the competence of public agencies to deal with coal production intelligently and responsibly. This intermixing of public and private institutions in fashioning a future scenario for coal development is characteristic of all the nation's energy problems and extends, consequently, to all the nation's energy sectors. The erosion of a clear distinction between public and private energy sectors underscores the spreading politicization of energy decisions once largely, or exclusively, within the private sector. Of the salient political factors in coal decisions, we will now briefly examine the public opinion climate, the existing political formula, and the particular importance of regulatory and distributive techniques before turning in later chapters to an extensive examination of substantive issues.

THE OPINION CLIMATE

Coal policy is evolving in an atmosphere of profound disagreement between the national administration and the public over America's current energy condition. No broadly shared public agreement exists about the reality of the energy crisis nor about the origins of those energy problems Americans do recognize. The average citizen is often ignorant of basic energy facts; among energy matters, coal issues are ordinarily so removed from public consciousness as to be almost invisible. It is a climate uncongenial to incisiveness and innovation in coal management, a public mood highly conducive to official reliance on old (or slightly renovated) political formulas for coal management.

The absence of a crisis atmosphere is especially important. A crisis mood bespeaks a coherence of public opinion, compelling incisive, often creative, governmental response. Crisis is the cue forcing issues high on the political agenda, imparting urgency to deliberations, and uniting officials otherwise divided by partisan loyalties into at least sharing one definition of the situation: it fosters openness to ideas that might be considered untenable under less dire circumstances. A crisis environment also hastens the pace of policy formulation by limiting policy options—choices to "do little," "go slow," or look to tradition, especially, will seem dubious counsel. Crisis also inhibits, though never eliminates, the protracted bargaining between public officials and the pluralistic organized publics active in major policy questions; policy evolved in a crisis atmosphere is often more effective in dealing with issues because officials feel less obligation to balance and mollify all important publics affected by a policy. As it now exists, the public mood forces upon the president and other national leaders the responsibility for proving the reality of the energy crisis, for advancing new energy policies, for shouldering the burden of proof on the merit of proposals, and generally for forcing political action. Without the appearance of crisis, partisan divisions in the government loom as larger impediments to policy formulation, interest groups must be granted their customary due in policy-making, and innovation seems less imperative. Elective officials may well suspect, in light of the public's current attitudes, that the reward for responding to policy proposals as if an emergency did exist would be almost minuscule at the polls.

The opinion polls generally suggest a number of significant public beliefs about energy resources. The first of these is *a widely held, stubborn conviction—unyielding in the face of contradictory evidence—that an "energy crisis" doesn't exist.* The public mood in April 1977 was especially important. President Carter had bluntly warned the nation in an unprecedented address that the grave imbalance between domestic energy reserves and consumption would produce a catastrophe unless quickly rectified. The United States was then only a few months past its worst sustained energy crisis since World War II; during January and February 1977 much of the eastern United States had been virtually paralyzed by natural gas shortages. Nonetheless, a nationwide poll in April revealed that an estimated 51 percent of the population was convinced the energy situation was only fairly or not at all serious.[11] In September 1977 public energy concern was further diminished. Then, according to another poll, only a third of the population believed energy problems were "as bad as the president said" and only 38 percent believed the nation's energy shortages "are real."[12] A sense of urgency about energy conservation is often absent even among those citizens who recognize that the United States has energy problems. A Federal Energy Administration (FEA) consumer study in December 1976 revealed the frequent poll finding that energy conservation was considered significant yet lagged far behind other concerns in public priorities—the case of an issue being publicly important, but not salient. Thus, while 56 percent of the public in the FEA study thought energy saving was important, it ranked behind family happiness, preventing crime, fighting inflation, and job security in public priorities.[13]

Secondly, *the frontier has never been passed in the public imagination.* Behind the public's indifference to energy problems lies a fundamental culture issue: Americans have not truly accepted the passing of the frontier. "The very essence of the frontier," notes historian David Potter in his study of American national character, "was its supply of unappropriated wealth."[14] It is questionable, he adds, whether any other factor "has exerted a more formative or more pervasive influence [upon American character] than the large measure of economic abundance which has so constantly been in evidence" in American society.[15] Long after urbanism and industrialization obliterated the geographic uniqueness of the frontier, it has persisted in public thought as an unspoken assumption that the material sufficiency it epitomized would always be available. This presumption of resources, flowing from the western cornucopia, not merely abundantly but extravagantly, has so colored the public imagination that it is understandable why Energy Secretary James Schlesinger should explain public indifference to energy scarcity in terms evoking Potter: "The American people still have not recognized the fact or the nature of limits. . . . It's hard to adjust to the closing of the American frontier and we still haven't adjusted."[16]

The third conclusion the polls suggest is that *few Americans accurately understand present national energy inventories.* Americans seem uninformed about the particular energy sources most vulnerable to depletion. Apparently less than half the population is aware that the United States must import petroleum. Americans generally know little about the magnitude of natural gas and coal resources, their location, and their importance in the energy problem. Public consumption patterns show scant sensitivity to prospective energy shortages. On the eve of the 1973-74 energy scarcity, characterized by a brief but major constriction of the gas supply that frequently produced long lines at gas stations and temporary gas shortages across the country, the nation was importing about 33 percent of its domestic petroleum consumption; three months after the president's energy message, this figure had climbed to 48 percent.

Fourth, *Americans, in large numbers, blame energy problems on corporate conspiracy or other institutional factors.* Many citizens interpret energy shortages as a consequence of oil and gas company policies rather than as a problem of diminishing reserves; in late 1977, for example, about 49 percent of the public expressed a conviction that "we are just being told there are shortages so oil and gas companies can charge higher prices."[17] This sort of poor man's populism misdefines both problem and solution because it directs policy, by implication, toward energy corporations with the intent of increasing presumably underproduced fuel reserves rather than toward conservation of resources by public controls over private sector energy utilization.

Another finding the polls point to is that *Americans have been consistently opposed to direct or indirect governmental regulation of energy consumption to achieve conservation.* Perhaps because the energy crisis appears to many citizens as a corporate contrivance or a fiction, the public seems to resist sacrifices (sacrifices only in an American sense) which might be imposed through public agencies to achieve more effective energy management. Rather, the public seems to believe in the efficacy of voluntary citizen fuel conservation. The 1976 FEA consumer poll, presented in Table 1.2, is particularly informative in this respect because it indicates public preferences among numerous conservation measures from voluntary to governmentally imposed.

A public long accustomed to unlimited energy supplies and addicted to energy-intensive conveniences will understandably resist energy conservation. So broad a public aversion to forced energy conservation seems to suggest, however, that voluntarism is an invitation to futility.

Finally the polls suggest that, *to the extent the public has any persistent attitudes about coal utilization, it appears to favor greater production in spite of alleged risks.* Public opinion polls customarily devote scant attention to coal policy issues, preferring to concentrate upon the broadly publicized petroleum and natural gas issues. When confronted with specific questions about coal

TABLE 1.2

Public Preferences among Energy Conservation Policies, December 1976

Governmental Action	Percent of Respondents Who	
	Favor	Oppose
Provide information on how to cut down on gas, oil, and electricity	96	3
Require that new cars, air conditioners, refrigerators, heaters, etc., have labels indicating what the yearly cost for energy to run them will be	91	6
Set energy conservation standards for buildings, cars, air conditioners, and other equipment that all builders and manufacturers are required to follow	85	8
Point out to people why it is their patriotic duty to cut down on their use of gas, oil, and electricity	80	16
Offer tax rebates to people who install extra insulation in their homes, buy storm windows, and do other things to cut down on energy use	75	21
Remove price regulations and rely on free competition to determine what the price of energy will be	47	39
Ration the amount of gas, oil, and electricity each family can use and let people decide for themselves where they will cut down	39	57
Limit the available supply of gas, oil, and electricity and let people work out for themselves how to meet their needs	38	56
Add a special tax that would make it more expensive to use energy	22	71

Source: U.S., Federal Energy Administration, *Consumers' Attitudes Knowledge and Behavior Regarding Energy Conservation* (Washington, D.C.: December 1976). Prepared for the Office of Energy Conservation and Environment, Marketing Office, by the Opinion Research Corporation.

utilization, however, public response generally favors increased consumption, even when the inherent environmental problems are made explicit. Thus, in late 1977 a majority of the public indicated it approved of both strip mining and a relaxation of air pollution control standards, if necessary, to increase coal consumption.[18]

Such a national temperament implies, beyond the absence of a crisis atmosphere, some important consequences for the substance of energy policy likely to emerge from Washington. First, the public temper is hostile to energy conservation. It is difficult for public agencies to impose meaningful restraints on resources, especially when such restraints fall heavily upon the private sector, without vigorously supportive majority opinions and mobilized group pressure across the nation. Second, the public mood is very compatible with the rapid expansion of indigenous resources, such as coal, which appear to extract few overt public costs; unlike a gasoline tax, or a rise in gasoline prices attributable to incentives for more petroleum exploration, the increase in public costs for coal production (primarily for pollution controls) is usually hidden from the consumer. Finally, this public sentiment is one which implicitly promotes old solutions to new problems.

This last issue affects coal as it does any other energy source. It is important to emphasize that neither the American public nor its officials have experience in the long-term management of scarce resources; the nation has no public philosophy for dealing with critical resource scarcities—except during actual wartime conditions—and, consequently, no set of established public preferences or behaviors to provide precedent or sanction for choice among currently debated alternatives. Except for fuel rationing during World War II—a sacrifice legitimated by emergency and borne with widespread expectation the war would be brief—Americans have no way to assess, except problematically, the social impact of sustained energy conservation. In this situation, when the public interest cannot plausibly be associated with any specific set of policy programs, there will be a tendency for policy-making to transpire, as much as possible, within the old and comfortable rules; the nature of the problem does not seem to warrant major structural or procedural changes in public institutions or in the rules that govern decision-making. Even though policy-making officials may agree that an unprecedented policy challenge exists (many officials now proclaim this ritualistically, in any case), they are still likely to fashion new policy with old tools and thereby lean heavily upon the past.

In any event, the realities of public opinion throw into sharp relief the importance of existing public arrangements for coal management, together with inherited historical experiences in coal utilization. These become the established political formula likely to be regarded as the most immediately relevant cue to officials in deciding how to handle future coal management.

THE POLITICAL FORMULA

Future coal policy will be written on no *tabula rasa*. In the United States a specific political formula has existed for determining how coal will be utilized on public and private lands—an amalgam of laws, bureaucratic structures, private associations, and traditional procedures which fashion these components into a cluster of policy-making elements. Additionally, a generalized political style, observable in governmental management of other resources, is likely to affect coal should it now become a strategic national resource. For convenience, we can illuminate these different factors by first examining prevailing public coal philosophy as it is expressed in law, governmental structures, and interest-group arrangements. Then we shall examine the political conventions which have more generally characterized strategic resource development since World War II.

"To Foster and Encourage Private Enterprise"

The frontier never died in the coal fields. Public coal philosophy has been, more than anything else, the political incarnation of the frontier ethic. Expressed in contemporary law and bureaucratic arrangements, this durable policy has been a robust expression of *laissez faire* economics softened only by a mild and tardy conservationist ethic. Its most explicit expression is in the arrangements characteristic of federal land management where most of the nation's coal reserves lie under federal guardianship.

"The brand of extreme individualism that necessarily characterized the frontier dominated our attitudes toward all resources during the nineteenth century," writes Stewart Udall. "It was an era when we raided the Indians, raided the continent—and raided the future."[19] There was a conviction dominating public attitudes, mirrored with great fidelity among governmental structures, that the national interest lay in opening all public lands—upon which resided originally most of the nation's extravagant natural abundance—to private use without much restraint. Federal resource policy from George Washington's time to the turn of the century largely presumed ". . . all land should be made private property as rapidly as possible, that new states should be settled with the utmost rapidity; with the general conviction that *laissez faire* is the best guide in deciding questions of conflict between government and citizen."[20] Throughout most of the nineteenth century, this developmental bias was expressed toward coal, as toward other minerals, by rapid governmental divestment of public domain with its rich coal seams into private hands.

The advent of the conservation movement at the turn of the century gradually forced the federal government to hold back these giveaways and to exercise greater stewardship over the remaining public lands. This newer public-land ethic and the political structures which resulted are especially important for coal because more than 58 percent of the nation's western coal reserves still rest under federal control, as Table 1.3 indicates.

Generally, federal guardianship of these resources was rationalized as an effort to foster multiple or balanced use which would ideally permit only such future private use of public resources as was compatible with good conservation

TABLE 1.3

Federal Ownership of Western Coal Lands
(millions of acres)

State	Federal Coal	Nonfederal Coal	Total	Percent Federal
Colorado	8.7	7.9	16.6	52
Montana	24.6	8.2	32.8	75
New Mexico	5.5	3.9	9.4	59
North Dakota	5.6	16.8	22.4	25
Utah	4.1	0.9	5.0	82
Wyoming	19.8	10.7	30.5	65
Total	68.3	48.4	116.7	58.5

Source: ICF Incorporated, *Energy & Economic Impacts of HR 13950: Final Report, Appendix* (September 1977), p. F–48. Report submitted to the Council on Environmental Quality and Environmental Protection Agency.

practices. In the case of coal, this produced a nebulous mandate to federal agencies to achieve a balanced use which in reality led to political arrangements largely congenial to coal utilization under terms indulgent of mining interests.

Institutional Arrangements

The governmental apparatus erected about coal in the public domain consists of several components: first, a legislative mandate requiring balanced use but giving priority to private utilization; second, the delegation of administrative responsibility for achieving this "balance" to the Department of the Interior which has been preoccupied with coal development, and with other goals sympathetic to private resource users—the familiar instance of a regulatory agency responding to the interest of the regulated; third, the appearance, elaboration, and institutionalization of strong formal and informal political ties between regulated resource users and the regulatory agency; and fourth, the gradual growth of a community of interest between regulators and regulated. By the middle of this century, these various arrangements matured into political infra-structures which organized major public agencies and private coal interests into a political alliance largely sensitive to developmental and distributive policies for coal.

The federal bias toward mineral production from the public domain is evident in the priorities for mineral development assigned to Interior by the Mining and Minerals Policy Act (1970) which directs the Secretary of Interior:

> . . . to foster and encourage private enterprise in (1) the development of economically sound and stable domestic mining, minerals, metals, and mineral reclamation industries; (2) the orderly and economic development of domestic mineral resources, reserves, and reclamation of metals and minerals to help assure satisfaction of industrial, security, and environmental needs; (3) mining, mineral, and metallurgical research, including the use and recycling of scrap to promote the wise and efficient use of natural and reclaimable mineral resources; and (4) the study and development of methods for the disposal, control and reclamation of mined land . . . [21]

The Department of the Interior, which had managed the public lands and virtually all their mineral reserves since its creation, is formally constrained by numerous congressional enactments, as well as its own regulations, to discharge these broad responsibilities in a manner protective of the public's interest in conservation. Interior's benign regard for the mineral industry, however, is legendary; the department is widely viewed throughout government as acutely sensitive to "coal people" and generally expressive of their viewpoint. Reformist secretaries have tried many times, with only sporadic success in the last several decades, to weaken this strong identification in order to introduce greater concern for conservationist sentiments.

The principal administrative procedure through which the department affects coal interests is the assignment of coal leases. This procedure in the past permitted the department great freedom to decide, subject to some broad statutory guidelines, what coal reserves would be explored and mined, what rentals and royalties would be payable to Washington and to the states within which the resources reside, and what regulatory controls applied. When interpreting these responsibilities, the department has usually deferred to the convenience of mining companies and speculators. Leases were often sold cheaply, frequently without competitive bidding, and without pressure for immediate use or mandatory deadlines for development (a practice permitting speculators to determine when production could begin on the most favorable market terms).[22] Thus, although the department was ostensibly minimizing the pressures for immediate coal production, it was ensuring that in the longer run the coal would be developed on the terms most satisfactory to mining entrepreneurs. This stance, together with the department's customary lassitude in imposing required environmental safeguards on the mining enterprises, largely served the coal industry. Indeed, in 1976 the House Committee on Interior and Insular Affairs complained that the department seemed to act "as if the lands were private property."[23]

Coal has often bound the outlook of the department to another major sector of the energy community. Electric utilities frequently own coal mines, or mines and utilities may both be constituents of a corporate conglomerate. Moreover, coal is a primary fuel for the electric generating industry. As Table 1.4 demonstrates, electric utilities are overwhelmingly the largest consumer of mined

TABLE 1.4

Consumer Use of Bituminous and Lignite Coal, by Amount and Percent of Total Consumption, 1974

	Amount (thousands of short tons)	Percent of Total U.S. Consumption
Electric utilities	392,551	70.4
Industrial[a]	93,158	16.7
Residential and commercial	6,792	1.2
All other[b]	64,754	11.6

[a]Coke and gas plants.
[b]Includes railroad fuel, mine consumption, and all other miscellaneous categories.
Source: U.S., Department of the Interior, Bureau of Mines, *Minerals Yearbook, 1974*, p. 395.

coal. The Department of the Interior often collaborates with states and private utilities to promote the development of large power generating complexes. For example, in building the sprawling Four Corners Project at Farmington, New Mexico (which provides power to major western metropolitan regions including Los Angeles and Tucson), the utility consortium responsible for operations was assisted by the department in providing water to the coal companies providing the fuel used for power generation. This collaborative intermingling of public and private power is one revelation of the complex interest structure which continually interacts with the department in policy formulation and implementation. This group structure embraces major private coal associations, related economic organizations, and other state and local governmental units that have become the department's clientele.

Interior's Clientele

The western states embrace almost all the public land within Interior's jurisdiction. They have always had a strong interest in Interior's administrative performance; indeed, so much of the land in some states like Nevada and Alaska is in the public domain that federal mineral policy is inevitably the dominant force in state mineral extraction decisions. As the Four Corners Project illustrates, the western states have demonstrated keen interest in coal mine development and, quite often, in electric utility expansion. While acutely mindful of the environmental risk in coal mining, the states have been attracted by the very substantial royalties now generated by mining, the taxes generated by new utility plants and services, and by the added growth potential offered by increasing power generation capacity at utility sites near mines. Public land policies affecting these western states, and particularly regulatory decisions formulated by Interior, are customarily surveyed and often modified through complex negotiations between Interior and the states.

Private associations daily involved in Interior's activities include such obvious interests as the coal mining companies, public and private utilities, petroleum companies owning mining operations, the railroads transporting coal, unions representing coal miners, and conservation groups. Of particular importance in this constellation of groups are the private associations that intermix, through membership rolls and organizational activities, the personnel of public agencies formulating coal policy and the clientele groups affected by the policy. Of these, the National Coal Association (NCA) and the National Minerals Congress (NMC) are especially effective in the political arena. The community of interest between the NMC, the NCA, and Interior was thrown into sharp relief by their cooperation in lobbying against proposed strip-mine regulations which eventually became the Strip Mining Control and Regulation Act of 1977. The NCA especially has achieved a reputation as the most authoritative voice for coal people in national politics.

Thus, while coal has been a laggard compared with other energy sources since World War II, through its abundance in the public domain it has precipitated a well-organized political structure combining federal, state, local, and private interests in an alliance generally protective of its interests. Even in the absence of past incentives for massive development, the developmental psychology still dominates the environment in which coal policy is made on the public lands.

Coal on Private Lands

Approximately 65 percent of the national coal reserves resides in private lands; more than 80 percent of the nation's current coal production originates from this source. Prior to the 1977 passage of the Strip Mining Control and Regulation Act this was a private domain largely immune to effective governmental control; rather, market forces had traditionally dictated rates and distribution of coal production.

Until recently, federal concern with the mining industry was confined almost exclusively to the enforcement of mine safety standards and to intermittent intervention in the industry's historically violent and continuously turbulent labor relations. The states had the responsibility for regulating as they saw fit the environmental hazards of deep-shaft mining and the visibly grave ecological degradation of surface-mining—a regulatory task, becoming more imperative especially as stripping rapidly swept the continent, that was commonly indifferently and often negligently exercised. In Appalachia, this failure was an expression of the enormous political and economic weight wielded by mining companies in state and local politics. Regulatory laws were passed and routinely violated; enforcement was so feeble as to be largely a charade. In the Midwest and Far West, states were somewhat more diligent, yet as late as 1976, on the eve of federal strip mine regulation, a congressional staff study concluded that state regulation was "in many cases, simply inadequate."[24]

In addition to the coal producers, the trade associations, and the governmental bureaucracies with some regulatory responsibility over coal production (the agencies concerned with working conditions, labor relations, and environmental impact) the most important remaining interests concerned with coal production on private land have been the railroads and the United Mine Workers (UMW). Once strongly organized under the legendary John L. Lewis and including approximately 540,000 members, in the last several decades the UMW has steadily lost membership as coal production declined and strip-mining (a less labor-intensive practice) replaced deep-shaft mines across the nation. For the last decade, the UMW has been torn by factional strife and wildcat strikes which demonstrate a lack of dependable control from the ostensive leadership structure. Politically, relations between the UMW and the mine operators have largely consisted of recurrent bargaining struggles, frequently erupting into strikes, over wages and working conditions, yet never undermining union support for increased coal utilization. This combination of adversary and collaborative stances toward management is guided by the UMW's understandable preoccupation with job security. Indeed, even under the fiery John L. Lewis, the UMW was so often accommodating to management arguments about the importance of protecting production—even at the cost of some union benefits—that critics frequently charged a sweetheart relationship existed between labor and management. Of more immediate importance, the UMW is generally a dependable ally of the coal industry in advocating increased production.

Railroad concern with coal policy is obvious and direct: the railroads have traditionally carried coal from mining sites to delivery points. Between 1970 and 1975, the percentage of coal shipped by rail in the United States rose from 68 percent to 77 percent.[25] During the coal boom early in this century, about 20 percent of all railroad freight was attributable to coal; in 1976 this had risen to approximately 25 percent of railroad freight cargo and, thus, coal represents a major component of service demand for an industry which has been suffering a severe decline since 1945. America's railroads regard prospective new coal production as a rare glimmer on a generally darkening economic horizon; to the president of the Chicago and Northwestern Railroad, President Carter's call for massive new coal utilization was "the great black hope for railroads."[26] Understandably, railroads have been a political ally of the coal industry and, more recently, one of the most vocal advocates of coal utilization as a solution to national energy ills.

TABLE 1.5

Percent of U.S. Bituminous and Lignite Coal Shipped by Rail, 1970-74

Year	Percent
1970	68.1
1971	69.2
1972	66.2
1973	67.1
1974	77.1

Source: U.S., Department of the Interior, Bureau of Mines, *Minerals Yearbook, 1974*, p. 59.

The configuration of political forces associated with coal mining on private land, together with those active in public land policy, roughly define the political factors most immediately active in current coal policy. As the political givens most immediately relevant to contemporary coal production, these elements will inevitably influence future coal policy. But a massive, short-term increase in coal production is likely to engage other political forces hitherto rather secondary in traditional coal management.

NEW MANAGEMENT STRATEGIES

A major short-term expansion of coal output within the United States would almost certainly alter the existing political context of coal policy in some predictable ways. It would require unparalleled governmental intervention in the coal sector to achieve the planning inherent in the massive substitution of coal for domestic petroleum and natural gas consumption. Specifically, it would require extensive regulatory arrangements for coal management and the elevation of coal to a strategic natural resource with high developmental priorities. These events, in turn, would introduce some important new political dynamics to coal production which ought to be considered in any assessment of future coal development's risks and rewards. These elements shall be examined in general terms before their impact is extensively discussed in subsequent chapters.

Regulatory Politics

The nation's coal mines have recently fallen within Washington's regulatory ambit through the Strip Mining Control and Reclamation Act (1977).

Future short-term increases in coal production will require far more extensive regulatory arrangements affecting the transportation and market demand for coal. This new federal regulatory intervention appears inevitable because complex economic and technical changes within the United States will be required to create conditions congenial to rapid coal conversion; these changes are unlikely, and probably impossible, without governmental management. President Carter's coal proposals, though they may not be reflected in substantive detail in the coal policies eventually enacted by Congress, do illustrate why coal conversion and regulatory programs are linked in energy planning. In April 1977 the president proposed to increase domestic coal production by 370 million tons a year until, by 1983, the nation would be consuming between 1 and 1.2 billion tons of coal yearly, or about two-thirds more than it now does.[27]

Achieving this huge conversion would require extensive and otherwise unanticipated technological changes in private industry; an estimated 10 percent of the then existing facilities and 44 percent of future industrial plants would have to convert from petroleum or natural gas to coal—all this outside the utility sector. To accomplish these conversions, the president proposed "a revised and simplified regulatory program" that would prohibit industry and utilities from burning natural gas and petroleum in new boilers and would require the then existing facilities with a coal-burning capability to convert from petroleum or natural gas. Exceptions would have to be made, noted the president, when environmental or economic circumstances warranted. This was deceptively simple language. It largely obscured the need for a very extensive, and largely new, regulatory structure, including both substantive laws and bureaucratic arrangements, to implement each of these procedures. Nor did it suggest the extent to which demand for other fuels in many other economic sectors might have to be manipulated in order to achieve these coal conversion objectives. Indeed, the Senate Committee on Interior and Insular Affairs suggests that coal development implies a web of federal managerial and regulatory programs of unparalleled dimensions:

> ... miscalculations as to the factors determining supply and demand could gravely upset coal production targets ... each of the factors ... are crucial to success; their *timing* and *interlocking* aspects make success in some, but not in others, insufficient, suggesting a requirement for "synchronization" which can be met probably only through early and continuous U.S. Government oversight and possibly involvement on an unprecedented scale.[28]

The principal result of such regulation, as in any other regulated economic sector, is to substitute administrative, legislative, and judicial determinations for market forces and entrepreneurial choice in determining much coal policy. In practical terms, some predictable consequences would include:

• *the frequent subordination of economic to political factors in regulatory determinations.* This will occur despite substantive economic standards or criteria written by Congress into regulatory programs, or developed by agencies themselves, to govern decisions. Regulatory agencies customarily have strong, and probably irresistible, incentives to formulate policy that represents compromises among influential participants in coal policy, to build political constituencies and to mediate among their memberships, to develop an agency ideology, and generally to behave like any other public bureaucracy. In this respect a recent Brookings Institution review of the Federal Power Commission's (FPC's) work reveals the situation common to most regulatory agencies. Quite often, notes the study, the commission "acted as though politically acceptable compromise might be the primary goal" and, its economic mandates notwithstanding, adopted the "tacit assumption . . . that the FPC is performing satisfactorily when its 'clients' . . . do not complain loudly."[29]

• *a high incidence of legislative intervention in the administration of the regulatory program.* Commenting on the congressional penchant for regulatory approaches to economic management, the current chairman of the President's Council of Economic Advisors (PCEA), Charles Schultze, suggests that one explanation lies in the vulnerability of regulatory agencies to legislative intervention. Legislators "have the opportunity to hedge their [regulatory] programs," Schultze notes, by intervening on behalf of interests for which they are watchdogs. A regulatory program also offers the legislator a "constant opportunity to second-guess administrators and to provide services for constituents through intervention. . . ."[30] In short, those who may otherwise suffer losses through regulatory proceedings can, whether justifiably or not, attempt to protect their interest through legislative mediation.

• *the displacement from private management to administrative or judicial agencies of considerable responsibility for determining appropriate production technologies, marketing strategies, and logistical arrangements in coal utilization.* The extent of this displacement is still speculation. At a minimum, achieving new coal production on the scale contemplated by most planners will require federal agencies to make managerial decisions about appropriate environmental safeguards, about which industries and plants are eligible for conversion, and about acceptability of coal pricing practices. Ostensively, these decisions would emanate from regulatory agencies. But, as experience with air and water pollution programs illustrates, final determinations will often belong to the courts as litigation becomes the last resort for losers on the administrative front. According to many economists and political scientists, this displacement may be economically and technologically pernicious. Critics of such procedures increasingly assert, on the basis of current federal regulatory performance, that administrative determinations on technical production matters "lag behind technological and economic change," lead to political logic and costly delay in getting private sector performance, and produce confusion and uncertainty in production sectors.[31] To the extent that these indictments apply to new governmental regulatory intervention in coal management, they are hidden costs which must be weighed in deciding the best coal production strategies.

To predict some or all of these consequences does not necessarily argue against regulation; the performance of the coal sector in any future national energy program might not be superior in the absence of regulation. The political implications of regulation will, however, necessarily be as much a consequence of coal conversion as of any other energy factor and will therefore be as important to shaping future energy policy as anything else.

Distributional Politics

Like a pumpkin transmuted into a pearl, coal has suddenly acquired enhanced market value. Moreover, it is now clothed with public interest. This invites the president and Congress, who write the ground rules for resource promotion, to approach coal with the same political calculus used over the last several decades to apportion other resources suddenly acquiring similar importance. In particular, persuasive evidence now seems to compel a large public investment in basic research, in technology development and dissemination, and in rapid production and utilization of the resource. In deciding how these specific procedures will be implemented, a political style that can be called distributional tends to dominate.

Essentially, the considerations which preoccupy political decision-makers in developing a resource such as coal are *distributional impacts* (gains and losses translated into constituency terms), the pork-barreling of expenditures, and the opportunity costs or tradeoffs in selecting one policy option over another. In characterizing this as a distributional approach and emphasizing its hold on official thinking, one need not assume that other considerations (environmental or ethical, for instance) are banished from political deliberations. Nonetheless, these distributional elements will be particularly prominent and deserving of special consideration. There is nothing, in any event, esoteric in these components of public choice; they are the common stuff of American political style familiar to anyone acquainted with American public institutions. Distributional impacts of policy, for example, are the balance sheets of policy impact measured in constituencies as the basic unit; the standard of valuation may be economic benefits or losses from specific policies, advancement of local, regional or state prestige, or, if the constituency is conceived as social or economic interests, the relative satisfaction of such interests with policy options.[32]

Treating a resource as political pork means regarding expenditures for resource development and utilization as an occasion for competition and collaboration among legislators in search of shares for constituencies. Those who make the allocation decisions (the White House, congressional committees and leaders) tend to regard the decisions as a form of political exchange; this exchange logic often claims equal attention, if not priority, alongside the substantive merits of expenditures for any particular program. Thus, in pork-barreling expenditures such considerations as the electoral impact of specific appropriations, the status of the legislators competing for appropriations, political debts owed and collectable, and *quid pro quo* may figure strongly in how decision-makers choose to divide resource expenditures. Legislative logrolling—that is, vote trading—is usually the significant voting strategy when legislators confront a spending bill for public works in support of resource development. One impact of pork-barreling resource development is a tendency for expenditures to surge upward and hold at high levels because this enlarges the pie to be divided among claimants for a slice. This may be particularly noticeable in the budget for research and development—the ubiquitous R&D expenditures almost inevitably accompanying governmental development of new technologies.

Less important at this point than the particular mode through which pork-barrel logic is expressed is the prevalence of this exchange philosophy in Congress. It can hardly be otherwise, given the realities of American political culture. All legislators feel a constant burden to provide the folks back home with a share of federal expenditures. The vitality of this tradition is revealed in a 1976 Louis Harris poll which compared congressional and public perceptions of the congressman's role. To no politician's surprise, a majority of both groups gave top position to the legislator's responsibility to see that "his district gets a fair share of government money and projects."[33] This obligation is so imperious that many vigorous environmental spokesmen in Congress otherwise critical of Washington's ecological callousness in constructing public works were highly annoyed and uncooperative when President Jimmy Carter proposed to deauthorize projects in their constituencies on grounds of adverse environmental impacts.[34]

Narrowly conceived, opportunity costs are the economic benefits one sacrifices by choosing one mode of investment over another—the potential gain one theoretically gives up in preference for another prospective reward. Opportunity costs, more broadly defined, are constantly calculated by political decision-makers, not purely in economic terms but usually in terms of constituency and other interests as well; these costs are compared in some manner when deciding whether to choose one policy strategy over another.

Calculating these costs is so common to public officials that it is usually described, in various ways, as the normal mode of decision-making. Of more immediate interest, in resource policy (as in most other policy domains) the opportunity costs are often assessed over fairly short time intervals and often hastily—even though the rhetoric and rationalizations of those defending particular policies may suggest differently. The reasons for this situation are readily apparent. To begin with, the long-range consequences of policy decisions are often extremely speculative because adequate information upon which to make long-range judgments is lacking. Further, elective officials especially are often vulnerable to the tyranny of the immediate. Election is always but a few years distant; elective officials need to concentrate upon near-term policy impacts because these will be most relevant to voters. Then too, officials often feel public problems are too urgent to permit long-term study and reflection. In addition, opportunity costs tend to be reckoned more in terms of tangible than intangible considerations—because, among other things, counting the costs in dollars, or number of people affected, or some other real value is more easily defended than policy choices based primarily on moral or ethical alternatives (particularly when such abstractions are amenable to differing definitions).

None of these elements in distributional politics is inherently incompatible with sound resource allocations but they may be an unwise standard for any particular decision or they may dominate so often as to effectively bias institutional processes heavily in one policy direction or another. In any event, they constitute a common, pervasive influence in public policy-making and for this reason must be seen as political standards against which policies are measured and options are selected.

SUMMARY: THE FUTURE AS HISTORY

Public agencies, like politicians, have a strong tendency toward conservatism in their approach to formulating new policies. They have powerful—but not irresistible—impulses to do tomorrow what they did today when political ground rules and institutional arrangements for policy-making have been durable—in short, when the political formulas of the past appear to work. Many of the assumptions about resource use, including those relevant to coal, are implicit and unarticulated; they are unexamined and therefore uncriticized, even as they exert a powerful if often subtle influence upon how a resource problem is defined and solutions are devised. Further, powerful institutional interests, both public and private, have matured and structured themselves about resource allocations; they tend to regard radical or innovative approaches with suspicion if not hostility. Also, public opinion often provides little justification for reexamining established political formulas for resource use, especially because strong, mobilized, broad majorities demanding a major policy change are slow to develop.

All these factors throw into sharp relief the importance of examining, as this initial chapter has done, the existing political context in which the nation faces questions of future coal development. There is, as we have observed, a bias to the existing arrangements. It may not eventually dominate future coal policy but it defines the present dispositions of the governmental structure and thus becomes an influence upon the formulation of specific policy decisions. Having identified some of these salient political philosophies, institutions, and actors now influencing coal policy, we can explore the implications of coal development with attention not only to the substantive choices but to the interaction between policy options and existing political arrangements in coal management in an attempt to assess the political costs and rewards of specific choices.

NOTES

1. A useful summary of the industry's political history may be found in David Howard Davis, *Energy Politics* (New York: St. Martin's, 1974), chapter 2.

2. U.S., Congress, Congressional Budget Office, *President Carter's Energy Proposals: A Perspective* (Washington, D.C.: Government Printing Office, July 25, 1977), hereafter cited as "CBO"; and U.S., Congress, Office of Technology Assessment, *Analysis of the Proposed National Energy Plan* (Washington, D.C.: Government Printing Office, August 1977), hereafter cited as "OTA."

3. OTA, p. 9.

4. U.S., Executive Office of the President, Energy Policy and Planning, *The National Energy Plan* (Washington, D.C.: Government Printing Office, April 1977), p. xii, hereafter cited as "NEP."

5. CBO, p. xiii.

6. Ibid., p. xiv.

7. U.S., Congress, Senate, Committee on Interior and Insular Affairs, *The Geopolitics of Energy*. Energy Publication no. 95-1, 95th Cong., 1st sess., January 1977.

8. CBO, pp. 49-50.

9. Amory B. Lovins, "Energy Strategy: The Road Not Taken?" *Foreign Affairs* 55 (1976): 65-96.

10. See, especially, Charles L. Schultze, "The Public Use of Private Interest," *Harper's*, May 1977, pp. 43-62 James Q. Wilson and Patricia Rachal, "Can the Government Regulate Itself?" *Public Interest* 46 (Winter 1977): 3-14; and William Lilley, III, and James C. Miller, III, "The New 'Social Regulation,'" *Public Interest* 47 (Spring 1977): 49-61.

11. *New York Times*, April 16, 1977.

12. *New York Times*, September 1, 1977.

13. U.S., Federal Energy Administration, *Consumers' Attitudes, Knowledge and Behavior Regarding Energy Conservation* (Washington, D.C.: Government Printing Office, December 1976), p. vi.

14. David M. Potter, *People of Plenty* (Chicago: University of Chicago Press, 1954), p. 160.

15. Ibid., p. 165.

16. *New York Times*, October 2, 1977. Quoted by James Reston.

17. *New York Times*, September 1, 1977.

18. Ibid.

19. Stewart L. Udall, *The Quiet Crisis* (New York: Avon, 1968), p. 80.

20. Benjamin Horace Hibbard, *A History of the Public Land Policies* (Madison, Wisc.: University of Wisconsin Press, 1965), p. 562.

21. Cited in Thomas B. Stoel, Jr., "Energy," in Erica L. Dolgin and Thomas G. P. Guilbert, eds., *Federal Environmental Law* (St. Paul, Minn.: West Publishing Co., 1974), p. 933.

22. On coal leasing policy, see U.S., Congress, House, Committee on Interior and Insular Affairs, *Oversight of Federal Coal Leasing: Oversight Hearings Before Subcommittee on Mines and Mining*.

23. Ibid., p. 39.

24. U.S., Congress, House, Committee on Interior and Insular Affairs, *Surface Mining Control and Reclamation Act of 1974: Report*, May 30, 1974, H. Rep. 93-1072, p. 61.

25. U.S. Department of the Interior, Bureau of Mines, *Minerals Yearbook, 1974*, p. 59.

26. *New York Times*, May 2, 1977.

27. NEP, chapter 1.

28. Senate Committee on Interior and Insular Affairs, op. cit., p. 68.

29. Stephen G. Breyer and Paul W. MacAvoy, *Energy Regulation by the Federal Power Commission* (Washington, D.C.: Brookings Institution, 1974), p. 3.

30. Schultze, op. cit., p. 60.

31. James Q. Wilson, "The Rise of the Bureaucratic State," *Public Interest* 41 (Fall 1975): 97.

32. See John A. Ferejohn, *Pork Barrel Politics* (Palo Alto, Calif.: Stanford University Press, 1974).

33. "Testimony of Louis Harris, President, Louis Harris and Associates, Inc., Before the House Commission on Administrative Review," mimeographed, February 3, 1977.

34. Cited in Jack Shepherd, "A New Environment at Interior," *New York Times Magazine*, May 8, 1977, p. 42.

2
The Rewards of Development

If the American earth is to yield new coal, for whom shall it be a harvest? Coal development is as much economic as energy policy. A powerful economic stimulus to selective U.S. production, social, and geographic sectors is inherent in new coal production. Political alignments on coal conversion hew, in good part, to the contours of economic interest; economic benefits are a convenient means for identifying the distributive rewards of prospective coal development and for clarifying political cleavages. Economics is commonly a major, and often the preemptive, criterion for policy evaluation in the political arena.

The economic rewards of conversion can be measured in terms of their immediate economic growth consequences: to what extent they encourage more production capacity, more market demand, higher employment, secondary industrial and commercial growth, and so forth. But the full extent of the distributional consequences has been partially concealed behind the metaphors commonly applied to coal development and consequently current evaluations are often unrealistically crabbed in time; a longer perspective, albeit more speculative, is desirable. Further, the magnitude and distribution of economic rewards are often so sensitive to public policy and governmental behavior that the interaction among the government, the political system, and the economic distributions from coal conversion needs continual illumination.

COAL CONVERSION: THE GENERIC POLICIES

Coal conversion is less a single policy than a mosaic of specific policies directed at various sectors of coal production and consumption. The number of specific policies currently debated is relatively few. An informative overview of these policies can be found in the composite coal conversion program advanced by President Carter in his 1977 NEP. This plan is not a blueprint for governmental coal policy but it succinctly identifies the policy options from which a mix of strategies will be selected.

The NEP pointed to six basic and largely interdependent approaches to increasing further U.S. coal production and consumption:[1]

● *substitution of coal for scarcer oil and natural gas in the "near term" defined as 1977-85.* Within this time frame, the NEP proposed to increase domestic coal production by roughly two-thirds; by 1985, this would mean an annual domestic coal consumption between 1 and 1.2 billion tons yearly. By most authoritative estimates, this represents the maximum attainable output under realistic constraints of time, technology, domestic fuel pricing, and logistical limitations.

● *tax incentives for utilities and industry to switch from oil and natural gas to coal.* A "graduated users' tax" between 1979-85 was suggested in the NEP to drive the cost of scarce fossil fuel to a place where coal would be more economically attractive to consumers. Tax rebates or credits would be offered to industrial and utility plants converting to coal.

● *a regulatory program to force coal conversion in new industrial utility plants.* The NEP proposed to "prohibit all new utility and industrial boilers from burning oil or natural gas, except under extraordinary circumstances," and to "prohibit the burning of oil and gas in new facilities other than boilers." Moreover, "existing facilities with coal burning capability would generally be prohibited from burning oil and gas" until, by 1990, "virtually no utilities would be permitted to burn natural gas."[2] This compulsory change is the source of most anticipated conservation of natural gas and oil from coal conversion; most of this saving would presumably arise in the industrial sector.

● *conversion without environmental penalties.* The NEP promised both abundant coal and clean air. A number of regulatory procedures, including particularly a requirement that the "best available control technology" be used in all coal-fired plants, were included to buttress this commitment.

● *a major new coal research and development program.* Public and private investment in such a program is considered, by most observers, critical to the success of conversion. The NEP noted: "Expanding future use of coal will depend in large part on the introduction of new technologies that permit it to be burned in an environmentally acceptable manner, in both power plants and factories, for electricity, for process steam, and for heat."[3] This implies a major governmental investment in developing a host of new technologies: air pollution control systems, fluidized bed combustion systems, coal "cleaning" systems, solvent-refined coal processes, low and high-BTU gasification systems, synthetic liquids technology, coal mining technology—in brief, the rapid evolution of a major new American energy technology.

● *achievement of a "long-term" transformation of the American energy economy so that beyond the year 2,000 renewable energy sources will predominate.* Ostensively, this is not a substantive element in coal conversion but is, in fact, most pertinent because it creates in the NEP a theoretical boundary on intensive coal use planning. The evolution of nonfossil energy technologies—geothermal, solar, biomass, and others—is generally advanced as the most economically desirable means to prevent long-term reliance on coal as a primary fuel.

There are inherent constraints within this set of issues which necessarily govern any national coal utilization plan. First, while it is possible to select options for implementing any specific proposal—for instance, different tax structures can be used to hasten industrial conversion—the broad strategies generally presume one another; it is difficult to purge any major policy from the conversion mix without placing in jeopardy the other policies. Second, it is very unlikely that the United States can achieve coal production much beyond the 1.2 billion tons per year to which the NEP is oriented. In practical terms this means that under the most congenial circumstances coal in 1985 would provide about 29 percent of U.S. energy requirements and reduce demand for oil by the equivalent of 2.4 million barrels per day.[4] Finally, the geographic dispersion of coal reserves and consumers determines in large measure where economic stimulants will be felt through national coal conversion. Opportunities for governmental intervention to manipulate the distribution of economic benefits lie preponderantly in future technology development.

The economic impacts can be illustrated by examining, in turn, the developmental and political implications of mine siting, coal transportation, new industrial and utility consumption, and future technology development.

MINE SITING

The greatest regional impact of increased coal production will be felt in Appalachia and the Rocky Mountain states.[5] National coal conversion would bring intensive, widespread coal mining into the Rocky Mountain states for the first time; a population explosion triggered by the vigorous stimulation of new coal production could dramatically alter the socioeconomic character of the region within a few years. Within a decade these states—now sparsely settled and predominantly agricultural—would be transformed into a new, populous, regional aggregate of coal-based state economies. Appalachia, already the most extensively coal-mined region in the United States, would experience a new proliferation of coal mines vastly exceeding in number (though not in output) any incremental mine growth elsewhere. Population would increase but the region's social character would not undergo the metamorphosis of the western states. Together, Appalachia and the Rocky Mountain states would become economically, socially, and politically the nation's new coal bloc.

The magnitude of new coal mine development within a specific state or geographic area depends upon many factors, such as prevailing environmental standards and the extent of industrial coal conversion; it is possible, however,

TABLE 2.1

Demonstrated Coal Reserves in the United States, 1974
(millions of tons)

Region	Estimated Reserves Strip Mine	Underground	Percent of Total Reserves
Appalachia			
Alabama	1,184	1,798	
Kentucky	7,354	18,185	
Maryland	146	902	
Ohio	3,654	17,423	
Pennsylvania	1,181	29,819	
Tennessee	319	667	
Virginia	679	2,971	
West Virginia	5,212	34,378	
Subtotal	19,729	106,143	29.0
Midwest			
Arkansas	263	402	
Illinois	12,223	53,442	
Indiana	1,674	8,949	
Iowa	—	2,885	
Kansas	1,388	—	
Michigan	1	118	
Missouri	3,414	6,074	
Oklahoma	434	860	
Subtotal	19,397	72,730	21.3
Rocky Mountain-			
Pacific Coast			
Alaska	7,399	4,246	
Arizona	350	—	
Colorado	870	13,999	
Montana	42,561	65,834	
New Mexico	2,258	2,137	
North Dakota	16,003	—	
South Dakota	428	—	
Utah	262	3,781	
Washington	508	1,446	
Wyoming	23,845	29,491	
Subtotal	94,484	120,934	49.7

Source: U.S., Department of the Interior, Bureau of Mines, *Minerals Yearbook, 1975*, p. 353.

to identify those states and regions where significant new mine sites will probably develop under a fairly ambitious conversion program to 1985. Table 2.1 indicates the demonstrated coal reserves among the states and aggregates these states into regional groups. The importance of western coal is apparent in this table, which indicates that approximately half the nation's reserves lie in the Great Plains or further west. Equally important are the data in Table 2.2, which indicate the estimated new mine sites that would be required by 1985 if the United States were to attain the maximum feasible coal production increase to 1.2 billion tons yearly; this estimate, one of several provided by the OTA, appears to be the most realistic in its assumptions about future production conditions. Table 2.2 indicates that new coal production, almost all surface-mined, would occur in twenty states. Although the largest number of new mines (177) would appear in Appalachia, the 91 new mines in the West would be far greater in production. Altogether, the OTA notes, the outbreak of new coal mining necessary to achieve the 1985 levels considered most desirable "has no antecedent in our history."[6]

In the national debate over coal development, discussions of new production have understandably focused on potentially adverse environmental impacts because they seem immediately ominous. Coal mine development, however, is also an economic stimulus in regions where it appears. Generally, new mining creates jobs (the extent depending on volume of production and whether mines are underground or strip), generates mining royalties to the affected state governments, creates a new tax base for local governments, spurs rapid urbanization with all its associated service industries, and sometimes leads to the appearance of ancillary industries that take advantage of the proximity of coal supplies. In the longer run, mine development, particularly when high-volume and sustained, can lead to the appearance of mammoth mine-mouth electric generating plants and the growth of transportation systems linking coal to its major markets. Since the western states are likely to experience these transformations most profoundly, it will be instructive to examine in more detail some of the implications.

More than 40 percent of the nation's coal reserves lie below eight western states: Colorado, Montana, New Mexico, North Dakota, South Dakota, Utah, Washington, and Wyoming. Because this coal is abundant, accessible through strip-mining, low in sulfur, and reasonably close to major markets (primarily midwestern utilities), it is likely to be heavily developed in the near future. A 1975 federal study of six western states most likely to experience new coal development suggests major growth impacts from significant new coal production would directly affect 99 communities.[7] For these, growth would probably be abrupt, extensive, and traumatic; more than a third are now settlements of less than 1,500 people and only 11 are larger than 5,000 individuals. Among all ten western states with significant coal reserves, the study estimates that population growth by 1985 would approach 300,000 people under assumptions of moderate new coal demand; this figure excludes any population increase

TABLE 2.2

Estimated New Mine Developments Required for
Maximum New Coal Production to 1985

Region	Low Estimate*
Appalachia	
Alabama	13
E. Kentucky	24
Kentucky-Tennessee-Virginia	30
W. Kentucky-Indiana	10
Ohio-Pennsylvania	10
SE. Ohio	5
NW. Pennsylvania	9
SW. Pennsylvania	12
Cent. Tennessee	5
NE. West Virginia	20
N. West Virginia	15
S. West Virginia	24
Subtotal	177
Midwest	
Arkansas-Oklahoma	3
Cent. Indiana-Illinois	10
S. Illinois	10
Cent. Illinois	3
N. Illinois-Indiana	5
Kansas-Missouri	1
N. Missouri	1
Texas	6
Subtotal	39
Rocky Mountain-Pacific Coast	
Arizona-Utah	—
Colorado-SW. Wyoming	5
NE. Colorado	—
SE. Colorado–New Mexico	—
Colorado-NW. New Mexico	21
W. Colorado	1
E. Montana	3
SE. Montana–NE. Wyoming	59
W. North Dakota	—
NW. South Dakota	1
NW. Utah	1
Washington	—
Subtotal	91
. Total	307

*Includes both surface and underground mines.

Source: U.S., Congress, Office of Technology Assessment, *Analysis of the Proposed National Energy Plan* (Washington, D.C.: GPO, August 1977), p. 234.

associated with the local appearance of new generating plants, transportation systems, or other industries in the wake of new mining.

In recent years, the federal government has undertaken a number of programs which mitigate many adverse impacts from such rapid growth and offer state governments new incentive for accepting more mining. These actions cast new mine development in a much more benign light than previously. In this instance, as in other aspects of western coal mining, the role of the federal government is a crucial matter. About 50 percent of all western coal reserves are directly controlled by Washington as part of the public domain and thus the federal conditions and constraints attached to this coal mining will profoundly affect the general character of future western coal mining. Two measures in particular have changed the economic status of western coal. The Federal Coal Leasing Amendments Act (1975) increased the royalties returned to states from new mineral leases on federal lands from 37.5 percent to 50 percent; additionally, royalties from surface-mined land rose from 5¢ per ton to at least 12.5 percent of the selling price. Congress directed that most of these increases be channeled to state subdivisions socially or economically impacted by mineral development.[8] In 1976 Congress passed the Federal Land Policy and Management Act, which gave state legislatures greater discretion in awarding other portions of mining royalties from federal leases to those subdivisions affected by mine development; related provisions created a federal loan fund to relieve states and their subdivisions from the costs associated with accommodating new mine development on federal lands. One effect of these new laws was to increase by about $44 million the amount of royalty money likely to be available to the Rocky Mountain states to alleviate the impact upon local communities of mining and other energy development in fiscal 1979. With the expected increase in coal mining on the public domain, federal royalties to the Rocky Mountain states with significant coal reserves will probably total between $1.5 and $2 billion between 1977–87.

Moderate federal mining royalties have been committed to the western states for decades but a new order of federal-state relationship seems likely to emerge from this recent combination of increased royalty rates, new federal incentives for state assistance to energy-affected local government, and the future growth of western mining. It seems to promise the emergence in energy planning of bureaucratic clientelism between the western energy states and Washington; state and local governments, suddenly recipients or distributors of large and growing federal subsidies, are likely to become an active constituency dependent upon, and committed to promoting, continued (and probably enhanced) grants for energy-related development.[9] Indeed, political leaders throughout the western energy states have already demonstrated a widespread conviction that federal assistance of this kind is both equitable and urgent. A statement in 1975 by the governors of nine Rocky Mountain states and Nebraska was prophetic: "Since the demand for development of Federal coal in the West is a result of national needs, then there is a corresponding national responsibility to insure adequate relief for environmental and socio-economic effects."[10] Increased federal funding of local energy development costs in the West would constitute, in effect, indirect subsidization in which industry (paying the royalties) and the rest of the nation (waiving its claim on such royalties for other purposes) would underwrite development costs otherwise borne by state and local governments. However equitable this arrangement may be, it is a powerful incentive to such development, removing one of the most apparent serious negative impacts associated with it.

New mine sites, particularly in the West, would have very significant secondary economic impacts. Mammoth mine-mouth electric generating plants, certain to increase if utilities are forced to convert to coal, would probably multiply further with the availability of new mining sites. New or existing transportation systems would appear to carry coal from new mines to new markets. New mines almost inevitably create complex physical infrastructures to utilize their products.

MINE-MOUTH GENERATION

Fiscal or regulatory programs to force coal conversion in new utility plants or to encourage conversion to coal in existing utility or industrial boilers will stimulate more "mine-mouth" installations. Especially in the West, the appearance of new mining sites in great abundance will add further incentives for this growth. "Mine-mouth generation" involves the building of (usually mammoth) electric utilities in close proximity to coal supplies. This assures generating plants a long-term, dependable fuel supply essential to their economic viability; it drastically reduces costs of fuel acquisition, thereby giving market advantages to many power companies. Mine-mouth generation, though sometimes impossible for logistical or economic reasons, is fairly common in the American power industry. In 1977 approximately one-fifth of the electric power generated in the United States originated from mine-mouth installations. Conservative estimates suggest that the power generated from coal at mine-mouth installations will increase from 5 percent of all coal-generated western energy in 1975 to 25 percent in 1986—a fivefold increase in a decade.[11]

The federal government is already stimulating mine-mouth development indirectly through current coal conversion programs. Acting under authority granted by the Energy Supply and Environmental Coordination Act (1974), the FEA has ordered 11 utilities to convert existing oil and natural gas boilers to coal; another 91 existing utility installations are being considered for similar orders. Anticipating the future in this, the utility industry is already engaged in a "massive voluntary switch from oil and natural gas to coal and nuclear energy for new electric generating facilities" which, the NCA states, will mean 250 new coal-fired plants by 1985.[12] As coal increasingly becomes a primary energy source for future utilities, mine-mouth generation will continue to appeal as a siting option well into the next century.

The growth of new mine-mouth generation would be particularly explosive in the western United States if new electric plants were compelled to utilize coal simultaneously with the appearance of the approximately 100 new mining sites expected to meet new coal demand until 1985; most of the desirable mine-mouth locations throughout the whole United States would be west of the Mississippi. A rough idea of the potential growth in western mine-mouth operations can be gained from data recently compiled by the U.S. Bureau of Mines. According to the bureau, about 59 new coal-fired plants are planned in the western United States. But another 105 are planned but unsited. Requiring that these unsited plants utilize coal would almost surely induce most of the builders, should they continue original plans, to seek mine-mouth locations.[13]

Federal efforts to force industrial coal conversion will also quicken the growth of mine-mouth electric power generation. The OTA suggests that many industries will probably choose electricity rather than coal when converting from other energy sources and in so doing push "on the electric utility industry the burden of coal conversion"—a burden which will presumably further increase utility interest in mine-mouth operations.[14] Mine-mouth generation is also an expedient solution to the economic and political problems which might otherwise arise if massive generating plants are sited near urban, suburban, or farming areas where public opposition is readily mobilized. Putting the plants at mine sites far removed from densely inhabited regions—that is, putting them in the Great Plains and Rocky Mountain regions especially—inflicts pollution and other undesirable externalities on empty land, small populations, and fragile but obscure ecological systems. Thus, the few and expendable often pay the social costs of electricity for the many.

Of what value are mine-mouth generating plants? They are powerful stimulants to economic growth in the regions they serve. They offer new electricity to sustain economic growth (and frequently have excess generating capacity to justify further regional growth); their operation requires generous manpower and physical resources beside coal. They produce new, and quite high, tax revenues for state and local governments. They attract investors. This impact is largely attributable to the scale upon which mine-mouth installations are constructed. A modern mine-mouth installation planned for federal land in Montana, but abandoned because of state opposition, suggests the magnitude of future installations. Costing $2.6 billion to construct and $1.1 billion per year to operate, it would have employed 7,359 workers by the year 2000. The plant would have required in a few years more than 76 million tons of coal annually—three times the total coal mined in Montana in 1975. In addition to its peak capacity of 900 megawatts, it would have daily produced 100,000 barrels of synthetic oil and one billion cubic feet of synthetic gas.[15] Coal-fired generating plants are among the better investment risks in the utility field, often producing 50 to 100 percent more revenue than oil or gas-fired installations. In short, there appears little doubt that sufficient capital is now available, and will continue to be, to underwrite new mine-mouth generating systems if opportunities arise. Given their capacity to generate jobs, public revenues, and energy that feeds economic growth near and far from the plant, these installations must be viewed, even with their negative aspects, as economically and politically attractive to many.

TRANSPORTATION

The nation's railroads, suffering an economic malaise more profound than the coal industry's, would seem certain to revive with a coal boom. Anticipating generous new revenue from future coal transportation, many railroad lines have already begun to invest in new rolling stock, to restore deteriorating track, and in general to contemplate an economic blessing pouring from the coal pits upon the railroads, almost uniquely among transportation systems. Under the most beneficent circumstances, estimates suggest that coal carloadings in 1985 might increase by more than 350 percent over the early 1970s.[16] In late 1977, railroad orders for new cars rose several hundred percent over other recent years; most of these orders came from western lines, where the biggest surge in demand is expected. The railroads have traditionally been the country's primary coal carriers, yet many analysts suggest that coal, despite appearances, may not be the black gold so touted in many railroad boardrooms.

One potential impediment to a railroad revival is logistical. A major portion of new coal demand is expected to originate through the conversion of existing and future industrial boilers from oil and gas. These new coal boilers, however, are likely to be widely dispersed geographically and to have comparatively small capacity. In some cases, no railroad tracks exist to supply such users. Then, too, the unit trains and other volume shipments of coal to a consumer by rail become economically viable at approximately 600,000 tons per year and financially attractive at 1 million tons per year. But even large industrial boilers rarely consume such volume yearly.[17] Thus, large and unresolved uncertainties exist concerning the economic feasibility of providing coal to new industrial users.

The railroads are considerably more apprehensive about the prospect of coal slurries. Many energy interests, including large energy conglomerates, are exploring or actively promoting slurry construction as an alternative to railroad transportation. Essentially, a coal slurry transports finely powdered coal, mixed in water to a thick suspension, over long distances under high pressure. The nation's only existing slurry, a 273-mile system linking coal mines in northeastern Arizona with consumers in southern Nevada, appears to be a promising prototype for much larger systems. In early 1977, eight slurry pipelines were planned in the western United States (where they are most appealing logistically and economically); these would have a combined length exceeding 4,700 miles and a capacity perhaps exceeding 94 million tons of coal a year.[18] More recently, an additional line has been proposed to float Wyoming coal 1,036 miles to White Bluffs, Arkansas. Theoretically, slurry lines can traverse the American continent as readily as gas or oil pipelines; they could provide high-volume, dependable, and apparently economically attractive coal supplies for consumers far from mines and remote from railroad tracks. The nation's railroads, resolutely resisting slurry development, have refused to grant the slurries the right-of-way across railroad tracks; lacking this, slurries are untenable. President Carter appeared to favor federal legislation granting slurry operators the right of eminent domain, which would neutralize railroad obstruction by giving slurry operators the legal ability to cross railroad rights-of-way. The slurries are also vehemently opposed by environmentalists and many western interests because great quantities of ground water are required to operate them. Nonetheless, slurries have considerable appeal to energy planners and public officials outside the West because they would seem to make possible an economically and logistically attractive way to get coal to new consumers.

NEW MARKETS

Expanded coal production requires expanded markets. Predictions concerning how much demand will develop, where, and when depend upon differing economic assumptions—hence the many varied scenarios offered by energy experts in projecting future national energy needs. Nonetheless, the broad economic impact of new coal utilization is clear. The new market must be primarily an expansion of the nation's present market, the industrial and electric utility sectors. Essentially, industry and utilities must enlarge existing coal consumption by converting, when possible, present gas and oil-fired production units to coal and by utilizing coal in future installations. The NEP proposed to accomplish this by a carrot-and-stick technique: a regulatory program would force industry to convert existing and new units, when possible, from gas and oil to coal while tax incentives would stimulate utilities to convert new or existing units. Regardless of how it is accomplished, increased coal utilization would have important selective impacts geographically and economically. It would become a powerful political and economic incentive to the growth of the power generating industry, would create new social infrastructures organized around coal utilization while magnifying the economic influence of such structures collectively, and would alter patterns of sectional coal dependency.

Coal conversion stimulates electric utility growth in several ways. We have observed that industrial planners, if forced to abandon gas and oil, may find electricity more attractive than coal as a primary fuel, hence industrial electric demand is likely to rise in the near future. Equally important, electric generating

TABLE 2.3

Estimated 1985 Coal Consumption under Moderate
Conversion Programs
(millions of tons)

	1974[a]	1985	Growth Rate[b] (Percent/Year)
Electric utilities	390	715	5.7
Household/commercial	11	5	−6.9
Industrial	94	151	4.4
Metallurgical	63	73	1.3
Synthetics	0	16	—
Exports	60	80	2.4
Total	618	1040	4.8

[a]Coal comsumption in 1974 was greater than production due to changes in inventory.

[b]Growth rates in this column reflect projected *average* annual growth between 1974 and 1985. Projected yearly growth figures may vary significantly from the estimate in some cases.

Source: U. S., Federal Energy Administration, *Energy Outlook, 1976* (Washington, D.C.: Government Printing Office, 1975), p. 21.

plants appear better able, technically and economically, than most industries to convert to coal use. "The electric utility sector represents the greatest potential for substituting coal for oil and gas between now and 1990," the FEA notes. "Synthetic fuels do not yet compete economically with natural gas and oil," it explains, and "increased coal consumption in the industrial sector is limited by the large scale required to employ coal economically."[19] From a price perspective, moreover, increased power demand makes coal even more attractive to electric utilities because they can purchase larger quantities at lower unit costs. Thus, the more industrial demand for electricity grows, the more attractive coal will seem to the power generating industry. The cumulative effect of these conditions is to make increased coal utilization and the expansion of the electric utility industry seem, to many planners, interdependent or at least simultaneously attractive. For these reasons, most energy planners assume that the largest increase in coal consumption within the next decade will be within the electric utility sector; this is evident in Table 2.3, which indicates likely shifts in coal consumption among different national economic sectors under conditions of moderate stimulation for new coal utilization (though not necessarily those in the NEP).

Another frequently ignored stimulus to electric utility growth, more difficult to estimate in magnitude, would be a growing national reaction against nuclear power. The economic effect of decreasing nuclear power utilization as an energy source is likely to be increased coal consumption because, in the words of the FEA, "coal consumption substitutes directly for nuclear power."[20] Estimates of future coal use, whether or not they are based on forced industrial and utility conversion, usually assume that presently planned new nuclear installations will come on line as expected at anticipated output. Should presently planned nuclear installations fail to materialize, the "missing" output would probably be recovered from coal and consequently coal and nuclear power would become a tradeoff in the immediate future.

The greatest increases in coal consumption are likely to occur in the East, Midwest, South, and Southwest, primarily due to the conversion of electric generating plants from oil and gas and the installation of coal-burning units in future new plants. The effect of this conversion on regional energy dependencies is particularly striking in the Southwest and New England, where coal consumption for utility and industrial use has traditionally been low. These regional transitions are suggested in Table 2.4, which describes expected shifts in energy dependence among national regions to 1985 based upon the assumption of moderate coal conversion. Clearly, New England and the west south-central states show the largest increase in utility coal consumption, from 7.4 to 26.8 percent and from 3 to 20.6 percent respectively. While some industrial coal conversion will occur in most regions, it will not significantly alter the general pattern of regional coal dependence described in Table 2.4.

The larger tale told in Table 2.4, as in most other estimates of future U.S. energy consumption, is that the nation's major regions will be much more coal-dependent within another decade than they are now. With the exception of the Pacific coast states, all national regions will apparently require significant coal supplies for commercial, residential, industrial, and utility power; for instance, according to the table, almost half the regions will be generating over half their total electric power by coal and another third will generate between one-fourth and one-half of their electric power in this manner.

With increasing coal consumption and the greater national dispersion of coal-dependent utilities and industrial plants, the economic infrastructure associated with coal can be expected to expand in size and social importance. This infrastructure involves the key productive segments of the coal industry—coal companies, unions, railroads, and perhaps slurry lines—and the major consuming sectors of the economy, primarily the electric utility industry and those industrial plants with large coal-fired boilers. Because electric utility construction is increasingly capital-intensive and less predictably profitable than previously, the search for investment capital to underwrite future plants is likely to be more aggressive within the utility industry. More private capital will be required, particularly to underwrite new coal-fired plants; this will apparently draw into the coal infrastructure a much larger investment by banks and other lending institutions with a consequent rise in the involvement of major financial institutions

TABLE 2.4

Estimated Percent Contribution from Each Fuel to Regional and Total U.S. Electricity Generation, 1960-85

	Coal			Oil/Gas			Nuclear			Hydro			Other		
	1960	1974	1985	1960	1974	1985	1960	1974	1985	1960	1974	1985	1960	1974	1985
New England	50.3	7.4	26.8	31.7	61.3	28.4	0.1	24.4	41.0	17.9	6.9	3.9	—	—	—
Middle Atlantic	69.3	42.7	47.9	18.5	36.2	13.6	0.2	8.5	29.9	12.0	12.6	7.3	—	—	1.2
East North Central	93.5	82.0	66.4	3.8	8.7	5.8	0.2	8.3	26.3	2.5	1.0	0.6	—	—	1.0
West North Central	40.3	54.4	70.1	46.9	27.2	4.9	—	7.7	17.2	12.6	10.7	7.7	0.2	—	—
South Atlantic	66.3	54.9	52.6	20.2	32.5	10.3	—	7.4	32.0	13.5	5.2	7.3	—	—	1.2
East South Central	74.5	76.5	50.8	5.5	5.4	4.5	—	3.6	37.3	20.0	14.5	7.4	—	—	—
West South Central	—	3.0	20.6	95.7	92.6	55.3	—	0.2	22.8	4.3	4.2	1.4	—	—	—
Mountain	11.8	46.3	48.7	36.6	23.2	16.9	—	—	14.9	51.6	30.5	15.2	—	—	3.7
Pacific	—	1.7	4.7	42.0	27.8	19.9	—	2.8	10.2	58.0	66.7	62.2	—	1.0	2.5
Nation	53.5	44.5	45.4	27.1	33.2	16.1	0.1	6.0	26.1	19.3	16.1	11.5	—	0.1	1.0

Source: U.S. Federal Energy Administration, *Energy Outlook, 1976* (Washington, D.C.: 1975), p. 241.

in the making of future coal policy by private firms and governmental bodies alike.

NEW TECHNOLOGIES:
SOFT AND HARD ENERGY FROM R&D

Nowhere are the economic distributions from coal conversion more amenable to direct political manipulation nor more likely to attract it than in the creation and refinement of new coal utilization technologies. The character of this research, moreover, will go far to send the nation down a harder or softer road to energy development. Perhaps unwittingly, the nation may be choosing very different energy futures through the implicit commitments it will make in the guise of R&D for coal utilization. For these reasons, federal R&D for coal utilization—normally a peripheral, if not obscure, issue in public coal policy debate—deserves careful examination.

Active federal participation in this R&D is inevitable. Creating new utilization technologies seems so essential a public interest, particularly since Washington has officially announced the energy crisis, that it is politically unavoidable. As a practical matter, experts generally assert that only Washington has resources in sufficiently generous measure to provide the massive infusion of capital and expertise necessary to increase coal utilization in the proportion necessary to curtail other fossil fuel uses. Further, the American public philosophy (or at least that version dear to the business community and persuasive to politicians) assumes that federal R&D funding should occur, as a matter of equitable cost sharing, in any new regulatory programs imposed on business and requiring for their implementation new or improved technologies. An expensive precedent has already been created in recent federal air and water pollution programs enacted since the early 1970s.* In any case, Congress commonly includes R&D funding in regulatory programs, where possible, for political expediency: it softens opposition from the regulated. The inclusion of federal R&D programs for coal utilization in President Carter's NEP was predictable. According to the NEP, a major federal contribution to future coal utilization would be "an expansion for the Government's coal research and development program, with special attention to demonstration of new technologies."[21] Any permutation of the NEP emerging from Congress will surely include a similar pledge.

---

*The funding of state and local waste treatment facilities under the Federal Water Pollution Control Act Amendments (1972) is a particularly graphic example of this federal financing.

Energy Pork

An important matter is the way in which this federal involvement affects the allocation of funds. Politicizing the R&D program implies that R&D becomes a new source of distributional benefits to the constituencies of the White House, Congress, and (to a lesser extent) administrative agencies. Thus R&D becomes "energy pork"—the newest in that line—and thereby subject to the political calculus so often governing the allocation of other federal expenditures amenable to pork-barreling. This implies a tendency for generous funding levels to be sustained or driven farther upward to assure new distributional sources for those in positions to award them. It would nurture a vested official interest in sustaining coal utilization research. Indeed, recent experience with the federal water pollution program strongly argues that once federal funding for technologies associated with resource development or control is committed, it is impossible to keep one-time commitments final, particularly when such funding becomes the reward that drives state, local, and private interests to conform with the law. Then too, when R&D becomes pork, specific allocations will often turn upon political considerations often irrelevant to or incompatible with more rational investment.

Sowing R&D also raises a hardy crop of dependent constituencies. A diverse array of interests—academic institutions, private corporations, research enterprises, state and local governments—commonly mobilize, under the inspiration of new research dollars, to protect or enhance their future access to them. This clientele will aggressively defend such R&D programs from cutbacks while imaginatively formulating justifications for fresh R&D. It is a constituency naturally allied with governmental institutions having a complementary interest in keeping the R&D flowing. In brief, coal utilization R&D will rapidly generate the political structures and behaviors common to other public R&D programs. More than just another illustration of funding politics, this sort of mobilization in energy research may profoundly affect the direction of future energy development.

The Hard or Soft Way?

A number of scientists have recently suggested that the energy crisis has poised the United States at the juncture of two different energy paths between which it will choose as it initiates national energy plans for the next few decades. The United States, they warn, is about to launch down the hard road resolutely, unintentionally, and perhaps irreversibly. Allegedly, this hard road is pernicious socially and environmentally but fortunately the nation may still choose a softer, more benevolent alternative. The hard/soft debate embraces many issues vastly transcending federal R&D spending for energy technologies alone. This spending, however, is among the most crucial of such issues. Essentially, some critics assert that the character of federal R&D spending on future energy technologies will become a major determinant of hard or soft directions. Long-term, massive investment in coal utilization technology is in this perspective a decisive move toward a hard energy future.

The "hard/soft" debate currently floats on a sea of contending speculations—hard and soft energy are themselves novel concepts—yet raises substantial questions about the politics of federal coal utilization research. In particular, it insists that the currently fashionable description of coal as a bridge fuel in federal energy planning may be a spurious metaphor in light of governmental long-range research plans. It also implies that the logic of federal R&D funding for coal utilization needs searching scrutiny, lest it become excessively committed to energy production at the expense of energy conservation—a common trend in other federal R&D for energy development.

Amory Lovins popularized the terms "hard" and "soft" to characterize the alternative energy paths he envisaged facing the nation.[22] Each of these paths contemplates a different scenario of energy development:

● *The hard energy scenario* would be a composite of future programs which are largely "an extrapolation from the past." The strategic features include an emphasis upon sustained growth of energy consumption and pre-occupation with minimizing oil imports; high priority for expanded coal production (largely through strip-mining and then conversion to electricity, gas, and liquid fuel), accelerated domestic oil production (from the Arctic and outer continental shelf), and rapid evolution of nuclear power for electricity generation (leading eventually to breeder reactors); and low priority for unconventional, nonfossil and nonnuclear energy.

● *The soft energy scenario* embraces less acceptance of the inevitability of a rapidly increasing energy consumption and a much higher electric power demand; considerable reliance on renewable energy sources including solar, geothermal and wind; the use of flexible "low technology" energy production—that is, largely existing, relatively inexpensive technologies capable of decentralization (such as district heating schemes); and the matching of "energy quality with end-use needs."

Like other soft energy partisans, Lovins sees at the end of the hard energy path a malevolent environmental and social destination. Firstly, the environmental risks in massive new coal, oil, and nuclear energy production are allegedly enormous and insufficiently balanced by reliably effective control technologies. Lovins is especially horrified at the prospect of a rapidly proliferating and international nuclear technology. Additionally, soft-liners emphasize the great inefficiency in energy conversion from fossil fuel to electricity—a waste in billions of BTUs annually.

But it is the economic and political consequences they prophecy, not the technical problems, that cause the greatest apprehension among opponents of hard technologies. Lovins predicts a technocratic nightmare: a pervasive corporate energy structure, spawned by hundreds of centralized energy complexes, interlocking with the governmental apparatus to siphon national resources relentlessly into further hard energy production. Hard energy, asserts Lovins, will require centralized management, heavy capital and labor investment, special market considerations, and preemptive rights to raw materials that will be obtained only by governmental concessions—all leading eventually to a corporate state in which energy institutions and the government work collaboratively. The social costs will be "subsidies, $100 billion bailouts, oligopolies, regulations, nationalization, eminent domain . . . "[23] In the new hard energy society the public welfare will be linked to the success of the required high technologies. Government will so securely insulate these technologies and their economic, legal, and political support systems that other social sectors will be increasingly denied available discretionary investment capital in order to nourish the energy complexes.

One need not accept the inevitability of such Orwellian tribulations to appreciate the immediate relevance of the hard/soft issue for coal. Essentially, soft energy advocates believe that current coal utilization policy will itself shape priorities for energy use in the long term and thus will push the United States in a hard or soft energy direction, whatever the consequences.

Coal and the Hard Road

Lovins has asserted that a massive governmental investment in the long-term development of new coal utilization technologies will not only set the United States resolutely down the hard energy road but make it difficult to turn subsequently to softer options. He writes:

> The innovations required, both technical and social [for the soft path] compete directly and immediately with the incremental actions that constitute the hard energy path: fluidized beds vs. large coal gasification plants and coal-electric stations, efficient cars vs. offshore oil . . . congeneration vs. nuclear power. . . . the pattern of commitments of resources and time required for the hard energy path and the pervasive infrastructure which it accelerates gradually make the soft path less and less attainable.[24]

In moving toward a soft energy path, he observes, "it is above all the sophisticated use of coal, chiefly on a modest scale, that needs development."[25] He advocates "technical measures to permit the highly efficient use of this widely available fuel," including the development of supercritical gas extraction, flash hydrogenation, flash pyrolysis, panel bed filters, and "perhaps most promising, the fluidized-bed system for burning coal."[26] All these measures, relatively safe environmentally and economically modest, would require "only a temporary and modest (less than twofold at peak) expansion of mining, not requiring the enormous infrastructure and social impacts" implied if coal were to be used through the year 2 000 for electric power generation and other purposes that replace oil. Using coal in this "sophisticated" manner, Lovins concludes, would buy sufficient time for the national development of softer, newer technologies to gradually replace primary reliance on fossil fuels after the year 2 000.

But a different approach could set the United States on the hard energy course. A heavy emphasis on the development of synthetic fuels from coal, principally through coal gasification and liquefaction, would increase coal dependency in the economy and hence coal utilization would simultaneously climb and magnify in economic importance. Federal encouragement of larger, more coal-dependent electric generating complexes, especially through the financing of new technology development compatible with such coal utilization, would also drive up coal demand, spur the expansion of electric generating capacity, and thereby accelerate national dependence upon coal rather than upon softer non-fossil fuels. Especially if federal R&D funding aimed at long-range, expanding coal utilization is combined with regulatory and fiscal incentives for the growth of the electric power and nuclear power industries, the effect would be a very substantial "sunk cost" by private and public agencies in hard technologies, which in turn diminishes the incentive to turn toward softer alternatives.

At this point one needs also to consider American political style—a matter only indirectly touched by Lovins. The powerful political attractions in "hard" energy development may well prove irresistible to public officials. Richard Gordon has pointed out that creating coal utilization technologies would generate a profusion of pork-barrel projects. He notes particularly that the new coal markets needed to increase national coal production will probably require the synthesis from coal of close substitutes for gasoline, light fuel oils, and methane. "Most [federal] proposals for increased coal use," he observes, "have centered around multi-million dollar programs to develop technologies for producing synthesized fuels from coal."[27] Additionally, hard technologies are labor-intensive. Creating new jobs is perennially fashionable in Washington. It implies new employment among many legislative constituencies. It is one common cure for stagnant local economies and a mechanism for deflating national unemployment rates, the bane of all administrations since the early 1970s. Estimates of the labor needs required to meet the energy independence aims of President Ford's 1975 state of the union message—a program not significantly different from President Carter's energy plan in its short-term goals—suggest that building the necessary energy-extracting-and-producing installations would require nearly

100,000 engineers, 420,000 craftspeople, and perhaps 140,000 laborers; indirect labor might be half again as great.

In comparison, soft technologies generally require much less manpower and capital investment. Public officials currently genuflect in the direction of nonfossil energy production and the media widely expose experimental programs, but actual governmental R&D proposals allocate comparatively miserly amounts to soft energy when compared with hard technologies. The OTA found this a major deficiency in the NEP and warned:

> ... the speed with which [new and renewable] energy resources are developed will depend on the commitments now made to research and development. The Plan does not commit the United States to the full range of incentives that are available for accelerating development of new technologies, including subsidies for private research.[28]

Allen Hammond, research news editor of *Science* magazine, was blunter: the NEP, he concluded, was "deadly hard" in its technology commitments.[29] Short-term federal commitments to hard energy development, of which coal utilization would be a major component, need to be carefully weighed for more than the near-term consequences.

A number of observers have warned that short-term energy planning becomes long-term as present choices preclude future options seemingly available now. This is certainly what the OTA had in mind when it noted that the NEP did "not address the question of whether planned changes in the United States energy patterns between now and 1985 will strengthen or weaken the base on which longer range development will take place."[30] Specifically, the near-term commitment of many billions of dollars in research to stimulate new coal utilization, the imposition upon industry and utilities of requirements for coal conversion, and the mobilization of the coal and transportation industries in the wake of prosperity born of a coal boom may create a coal utilization infrastructure powerfully and successfully resistant to major new energy conversions with the economic dislocations that might follow. Moreover, some economists have suggested that the heavy capitalization necessary to operate the greatly expanded hard technologies contemplated in the near term by most energy planners would dissuade banks and other lending institutions from any encouragement of competitive technologies for several decades. As these considerations suggest, it is more the political and economic obstacles than any inherent technical problems that would seem most to threaten the development of long-range soft energy production when present hard energy strategies are vigorously pursued in the near term.

CONCLUSIONS

Coal development is clearly more than an energy issue. It implies enormous potential economic stimulants to the nation, a pervasive transformation in the character of regional economies, the possible emergence of a new regional resource politics, and the creation or magnification of influence among selective energy production sectors. Ultimately it may turn the United States massively, and possibly irreversibly, toward an indefinite future of more fossil fuel utilization at the expense of alternate technologies. Looking only at the social benefits of new coal utilization—and this primarily through an economic perspective— some conclusions seem reasonable:

●American political sytle, particularly the political culture of government, is so congenial to the social distributions inherent in massive new coal utilization as to make the program seem almost inevitable.

●New coal development on the scale now proposed by most energy planners for the near term—until 1985—will create an economic transformation among Rocky Mountain states, turning them for the first time into coal-based economies. It will also mean the further growth of the electric power, railroad, and mining industries.

●New coal utilization, on the magnitude proposed by most energy planners, would likely create, for the first time, a large and *growing* economic sector committed to coal utilization. This sector would embrace the coal mining companies, railroads and other coal transporters, the electric power industry, and those other interests creating or producing new coal utilization technologies. This new economic infrastructure would probably become a major political force, particularly if it is linked collaboratively with political forces in the heavily coal-producing states.

●The only source likely to a countervailing force to massive new coal utilization is nuclear power. Present economic problems in the nuclear power industry, however, when combined with environmentalist opposition to "nukes," diminish the attraction of nuclear energy to the point where coal has no effective substitute in the near term.

●Governmental planning for future coal utilization seems to avoid questioning the wisdom or accuracy of projected future growth rates; rather, it accepts increasing economic growth, extrapolating projected figures largely from past experience.

●Governmental funding of coal utilization research may have a profound influence on the future direction of national energy development. Governmental funding politics is heavily biased toward hard energy spending. This may prejudice the later development of large-scale soft energy technologies.

Were it not for the environmental consequences, increased coal production would be, politically as well as economically, almost undebatable. Indeed, it would almost inevitably climb for many decades. The environmental consequences of coal utilization are sufficiently grave, however, to create major opposition—in fact, the only politically effective opposition—to unrestrained national coal development. The environmental issue, in both ecological and political aspects, needs to be examined and weighed against the developmental benefits examined in this chapter in order to appreciate the character of the national choice involved in increased coal utilization. The following two chapters concern this environmental dimension of coal use.

NOTES

1. NEP, pp. xix-xx.

2. Ibid., p. xix.

3. Ibid., p. xii.

4. OTA, p. 157.

5. On the impact of coal development west of the Mississippi, see especially Northern Great Plains Resources Program, *Effects of Coal Development in the Northern Great Plains: A Review of Major Issues and Consequences at Different Rates of Development* (Washington, D.C.: U.S. Department of Agriculture, 1975); and U.S., Comptroller General, *Rocky Mountain Energy Resource Development: Status, Potential and Socioeconomic Issues* (Washington, D.C.: Government Printing Office, July 1977), hereafter cited as "GAO Western Study."

6. OTA, p. 233

7. GAO Western Study, pp. 31-57.

8. Ibid.

9. The historic development of "bureaucratic clientelism" is briefly but informatively traced in James Q. Wilson, "The Rise of the Bureaucratic State," *Public Interest* 41 (Fall 1975): 77–103.

10. GAO Western Study, p. 55

11. Stanford Research Institute, *A Western Regional Energy Development Study* (Washington, D.C.: Council on Environmental Quality, 1976), p. 46 (available through National Technical Information Service as Document No. PB-260 835).

12. National Coal Association, "Achieving the President's Goal For Increased Coal Production and Use," mimeographed, p. 2.

13. U.S., Department of the Interior, Bureau of Mines, *Projects to Expand Fuel Sources in Western States* (Washington, D.C.: Government Printing Office, September 1976), p. 7 (available through National Technical Information Service as Document No. PB-265 633).

14. OTA, p. 5.

15. *New York Times*, December 15, 1975.

16. *New York Times*, September 20, 1977.

17. OTA, pp. 131-33.

18. Department of Interior, op. cit., p. 4.

19. U.S., Federal Energy Administration, *Energy Outlook, 1976* (Washington, D.C.: Government Printing Office, 1975), p. 193.

20. Ibid., p. 193.

21. NEP, p. xx.

22. Amory Lovins, "Energy Strategy: The Road Not Taken?" *Foreign Affairs* 55 (October 1976): 65-96.

23. Ibid., p. 91.

24. Ibid., p. 86.

25. Ibid., p. 84.

26. Ibid.

27. Richard L. Gordon, "Coal—The Swing Fuel," in Robert J. Kalter and William A. Voegely, eds., *Energy Supply and Government Policy* (Ithaca, N.Y.: Cornell University Press, 1976), p. 197.

28. OTA, p. 6.

29. *New York Times*, August 28, 1977.

30. OTA, p. 2.

3
The Costs of Coal: Ecology and Regulatory Politics

In the social calculus of coal policy, a single equation summarizes the short-term risks of new coal development: coal production policy is environmental policy. So direct and pervasive is coal's impact upon environmental quality that it is impossible to formulate a national coal production program without implicitly declaring the nation's environmental protection commitments as well. Coal production is related to environmental quality so consistently that no aspect of coal production leaves the environment unaffected.

Coal production and ecology are specifically related in important ways. Wherever coal is heavily utilized, grave and possibly irreversible environmental deterioration is possible; modern control technologies—often unproven—can never preclude the possibility of such damage. Consequently, one obvious cost to coal utilization can be reckoned in terms of the probable adverse environmental impacts directly attributable to its use. Moreover, there are secondary impacts: the physical and human resources required to support new coal production are, quite frequently, diverted from other economic sectors. Equally important, if more subtle, is the "cost" in terms of the air and water quality standards the nation may have to abandon to have its coal. Current development goals and national air quality standards especially are competitive; decisions in one domain will directly constrain those in the other. Thus, sharply escalating coal production will probably mean far fewer undegraded air quality regions than otherwise. Conversely, environmental quality commitments will largely determine the volume of coal, the proportion of mined eastern vs. western coal, and the costs of coal-generated energy to consumers.

Public officials have been loath to admit a possible incompatibility between coal production and existing air or water quality standards. Most have ignored the issue, as if to exorcise it with silence. Most often, they assert (like President Carter, who promised the nation could achieve his energy goals "without endangering the public health or degrading the environment") that environmental sacrifices can largely be vanquished through new regulatory programs to control ecological devastation from coal utilization.[1] Yet introducing new regulatory procedures for controlling coal's ecological impact, whatever else it

50

may produce, will involve consideration of other social costs. Further, it creates a new political structure for coal management that will powerfully shape coal's future development. One needs, in short, to calculate the costs of coal production in terms of both environmental impacts and the social impacts of another regulatory program.

REGULATORY POLITICS: AN OVERVIEW

It has become almost habitual for the Congress to prescribe regulatory programs to cure what it perceives to be chronically severe misallocations in the private economic sector. Beginning with the Interstate Commerce Commission in 1887, regulatory programs have proliferated until today more than 70 regulatory agencies exercise authority over most major economic sectors; over 50 of these programs have appeared since 1960. Customarily, these regulatory structures are a response to mobilized pressure groups representing major consumer sectors (farmers, labor, environmentalists) who assert that the public interest lies in correcting the inequities of the market that work to their disadvantage. A new federal agency is customarily ordained to enforce legislatively mandated constraints upon the behavior of the errant economic interests; to implement the mandate, discretionary judgment based upon legislative standards is permitted the regulatory bureaucracy.

The regulatory programs affecting new coal production will resemble in broad conception those created through the Clean Air Amendments (1970) and the Federal Water Pollution Control Act Amendments (1972) and enforced by the Environmental Protection Agency (EPA), created in 1972. These programs, somewhat different from older regulatory procedures because they control production processes in the economy rather than prices and market competition, assume a "standards and enforcement" approach to regulation.[2] Essentially, the programs create statutory standards for environmental quality, ordain what technical procedures must be utilized by polluters to conform with the standards (or what criterion must be used for procedures), empower specific regulatory agencies to elaborate and enforce both standards and control procedures, and attach penalties for noncompliance. Future coal production would be affected directly by the Clean Air Amendments and by the newer Strip Mine Control and Reclamation Act (1977), which also adopts this standards and enforcement logic.

After almost 75 years' experience with regulatory programs, it is possible to see their disadvantages with considerable clarity. Many problems are recurrent and virtually predictable; a recent study of the FPC notes these "lie in the structure of regulation itself" rather than in the substance of any particular regulatory program.[3] These problems need brief illumination in order to provide a context in which to examine specific aspects of environmental control associated with coal production.

Federal regulatory programs tend to have several liabilities. First, they usually fail to attain their objectives; frequently they badly fall short of statutory goals. This suboptimal performance often seems to "spring in large part from the practical difficulties involved in having regulators make complex managerial decisions or in finding incentives for private managers that would lead them to make decisions more to the regulator's liking."[4] According to this logic, the congressional attempt to substitute administrative judgment for the entrepreneur's often fails because the technical expertise and professional experience of administrators is inadequate and economic incentives for compliance on the part of those regulated are absent. Regulatory programs also thrust upon federal agencies a responsibility to regulate other federal entities or to control state and local governmental agencies. This is a common procedure in environmental control, where federal agencies such as the EPA enforce environmental pollution standards through oversight of state pollution control programs and agencies. Federal control of state and local agencies—and, more broadly, control of one governmental agency by another—often fails because governmental agencies enjoy an ability to resist regulation that private institutions do not possess. James Q. Wilson and Patricia Rachel suggest that this situation arises because a government agency operates "in a milieu of politically supervised autonomy." In such a circumstance few agencies can easily constrain one another:

> . . . each agency defends or enhances itself by mobilizing allies elsewhere in government—in the legislature, in other agencies, within the executive offices—who share a stake, material or ideological, in the agency's well being. No elected representative, no governor or President, no Congressional committee chairman, can allow the competition among public agencies to be settled, so to speak, *mano a mano*. Such an unsupervised and unmediated struggle could only reduce or deny the authority of political superiors. Their power depends on their willingness to exercise it, and this means their willingness to become involved to help define and protect the domains of valued subordinates.[5]

Since it is predictable that almost any governmental agency within the regulatory jurisdiction of another will have its political guardians, regulatory demands are frequently evaded or badly compromised in settlements reached by those protecting the autonomy of the regulated interest.

Regulatory agencies are often so enmeshed in writing complex regulations and in defining in detail the performance expected of the regulated that the regulatory process becomes glacial, cumbersome, confusing, and widely evaded. To implement its new water pollution control responsibilities, for instance, the EPA in mid-1976 had promulgated some 492 different effluent guidelines and had issued 45,000 individual plant permits. But in April 1976 the EPA had to withdraw all guidelines for organic chemical control because of a court challenge. Requests for an administrative hearing existed on fully a tenth of all issued permits. According to the GAO, widespread deviation from final agency guidelines existed in permit writing and industrial evasion of the permit conditions was common.[6] The EPA experience differs only in detail from that of most other regulatory agencies. The massive proliferation of rules with resulting enforcement and implementation problems seems inherent in customary regulatory approaches.

Regulatory agencies also commonly experience "bureaucratic clientelism" and discretionary authority—both imparting their own political dynamic to agency operations. Bureaucratic clientelism is the tendency of state and local governmental entities to develop close and dependent political ties with Washington's regulatory officers.[7] These subordinate agencies may depend upon federal agencies for funding; they may be continually monitored by Washington or may collaborate with federal agencies in various substantive programs. Whatever the reason, bureaucratic clientelism means that state and local governmental units become an active constituency of federal agencies. Federal administrators, needing the collaboration of this constituency and linked to it by formal and informal political ties, often formulate policy through extensive consultation and negotiations with these state and local units. Often, federal administrative policy reflects more the settlements reached through this consultation than decisions reached by other criteria. From a regulatory viewpoint, the problem in this arrangement is that federal regulatory decisions issuing from Washington agencies may reflect more deference to clientele viewpoints than to the substantive purposes of regulatory legislation—indeed, the two may be quite incompatible.

Administrative discretion is essential, and often constructive. It permits regulatory officials to exercise flexibility and judgment when applying customarily broad statutory mandates to specific regulatory issues. With growing federal regulation of highly technical industrial processes and complex economic sectors—a situation especially common to environmental regulation—this discretionary latitude has further widened. Because Congress is unwilling to encumber such legislation with enormous technical detail (which it is seldom competent to write anyway), it has chosen instead to defer to administrative expertise; regulatory agencies are expected to flesh out the law's generalities and relate them to regulated interests with technical appropriateness. Discretionary authority, however, promotes a highly politicized administrative environment. Administrators are vulnerable to intense, competitive pressure to exercise discretionary authority to the advantage of one or another of those with a stake in the outcome. Discretion can easily be utilized (perhaps unintentionally) to subvert the intent of regulatory programs, to defer to a regulatory agency's clientele at the expense of more legitimate concerns, to avoid or suppress issues—in short, to act in a multitude of ways hostile to regulatory objectives. All regulatory agencies have exercised their discretionary authority to the detriment of their regulatory mandates often enough to provoke considerable distrust of such discretion among partisans of regulatory programs. Even though inevitable, those points in the administrative process where regulatory discretion is exercised should be viewed as particularly vulnerable to pressures disruptive of regulatory intent.

Such ills mean that all traditional regulatory programs are freighted with liabilities, in varying degree, that must be reckoned as a political cost to their introduction into an economic sector. Nonetheless, the fabric of an environmental policy for coal will largely be woven on bureaucratic looms; these regulatory procedures are the principal strategy for preventing or ameliorating the ecological devastation always menacing in coal. Given these realities, an important consideration is the extent to which the specific regulatory arrangements promote or inhibit such costs. An answer will be more apparent by examining, in turn, arrangements for regulating strip mining and for controlling coal's air pollution—the two principal ecological issues tied to coal.

STRIP MINING

No technique of fossil-fuel recovery inflicts a more violent, pervasive, or devastating impact upon the land than surface mining for coal.[8] Without rigorous controls, stripping leaves the land a ravaged and often unrecoverable waste. Unregulated stripping deranges surface hydrology, creates diffuse acid mine drainage which often contaminates large aquatic systems, destroys all vegetative cover, creates heavy erosion and sedimentation of stream beds, destroys animal habitats, disrupts underground hydrology, and creates unstable "spoil" banks that menace public land and safety. Uncontrolled stripping so viciously disfigured Appalachia's wooded hills and hollows that its outraged people were the first to demand controls on the practice. There is little disagreement: uncontrolled stripping is an environmental catastrophe.

Stripping, now the common form of coal recovery, will increase still further with new coal demand. Strip-mine output has steadily climbed from 31 percent of all coal production in 1960 to 54 percent in 1975; as Table 3.1 indicates, all but one of the coal-producing states are now stripped to some extent.[9] Stripping will so extensively scour the coal fields in 1985 that three of every four tons of coal will then be surface-mined. Given this intimate association between coal production and stripping, it is obvious why environmental devastation looms so large as a risk in future coal production. While all coal producing areas would be heavily stripped to satisfy massive new coal needs, surface-mining poses particularly acute and unique problems for the western states.

TABLE 3.1

Production of Bituminous and Lignite Coal in the United States, by Type of Mining, 1974
(thousands of short tons)

State	Underground	Strip[a]	Percent Strip Mined
Alabama	7,053	12,771	64.4
Alaska	—	700	100.0
Arizona	—	6,488	100.0
Arkansas	—	455	100.0
Colorado	3,260	3,636	52.7
Illinois	31,256	26,960	46.3
Indiana	139	23,587	99.4
Iowa	379	211	35.8
Kansas	—	718	100.0
Kentucky	63,497	73,700	53.7
Maryland	90	2,247	96.1
Missouri	—	4,623	100.0
Montana	—	14,106	100.0
New Mexico	529	8,864	94.4
North Dakota[b]	—	7,463	100.0
Ohio	14,365	31,045	68.4
Oklahoma	—	2,356	100.0
Pennsylvania	42,249	38,213	47.5
Tennessee	3,106	4,435	58.8
Texas[b]	—	7,684	100.0
Utah	5,858	—	0.0
Virginia	22,767	11,559	33.7
Washington	15	3,898	99.6
West Virginia	82,220	20,243	19.8
Wyoming	526	20,176	97.5

[a]Includes auger and strip-augur mining.
[b]Lignite.
Source: U.S. Department of the Interior, Bureau of Mines, *Minerals Yearbook, 1974*, p. 357.

In an effort to minimize the environmental impact of stripping Congress passed in 1977 the Strip Mining Control and Reclamation Act. Long overdue and admirable in its intent, the act creates a new regulatory structure which will be, for better or worse, the principal bulwark against the environmental devastation otherwise implicit in coal surface mining. After summarizing briefly the act's major substantive provisions, it will be helpful to examine more closely its possible impact, particularly in the West, in order to appreciate its importance.

The Strip Mining Control and Reclamation Act

Broad in its jurisdiction, comprehensive and detailed in its substantive provisions, the Strip Mining Control and Reclamation Act, Public Law 95-87, signed by President Carter in April 1977, imposes upon strip mining a multitude of potential constraints long advocated by environmentalists. Following the logic of the standards and enforcement approach to regulation, the act:

• creates an Office of Surface Mining Reclamation and Enforcement within the Department of the Interior.

• establishes an "orphan bank" fund for the reclamation of abandoned mines. A reclamation fee of 35 cents per ton of surface-mined coal and 15 cents per ton of underground-mined coal is assessed against all nationally produced coal to provide these restoration revenues.

• requires that surface-mine operators keep waste materials off steep slopes, return mined lands to their approximate original contour, preserve topsoil for reclamation, stabilize and revegetate waste piles, minimize disturbances to water tables, and take other measures to minimize adverse environmental impacts.

• allows the states to assume jurisdiction over mines in their boundaries and to enforce federal standards under federal approval; states unable to implement regulatory programs, or failing to implement them after approval from Washington, will yield to federal enforcement.

• requires all mine operators to obtain a permit to be given only when operators demonstrate acceptable restoration plans and sufficient resources for the task.

• limits mining on alluvial valley floors to those lands where mining will not adversely affect agricultural production.

• authorizes the states and federal government to designate certain public or private lands as unsuitable for mining.

• permits states to regulate stripping on federal lands according to agreements reached between the two governments.

While the act falls somewhat short of the substantive goals desired by most ecologists (for instance, it permits mountaintop removal), it generally includes a very high proportion of the control procedures commonly advocated. Assessing the effect of these provisions requires attention to several matters. First, are these restoration provisions technically and economically feasible? Second, are the provisions politically enforceable? And what will be the spillover or secondary effects? In all these respects, the West is a special case.

The Science and Economics of Restoration

From the onset of the regulatory campaign, strip-mine companies have asserted that the cost of rigorous restoration would have a multitude of deleterious effects: it would drive companies out of business, greatly inflate production and consumption costs, and otherwise inflict a grievous economic burden on coal producers. Substantial evidence exists to the contrary, that the costs of compliance with the 1977 act will be quite sustainable by producer and consumer. "These cost impacts," notes a detailed study recently prepared by the Council on Environmental Quality (CEQ), "are not likely to significantly affect national coal production, coal consumption, coal prices, employment or electricity costs."[10] Even the most stringent restoration standards, according to another analysis, are "economically feasible using currently available equipment."[11] The economic issue can be confidently interred.

A far more disturbing problem is the feasibility of restoring western mined land to the ecological vitality required by the 1977 legislation. The states overlying most of the western coal enclose more than 2.5 million acres of strippable land. Much of this acreage would not be immediately mined. Very extensive new coal production could nonetheless gradually enlarge the mined territory to gargantuan proportions. The Northern Great Plains Resources Program (NGPRP) estimates, for instance, that substantial sustained new coal production would more than double currently mined land to 20 thousand acres by 1980, 70 thousand acres by 1985, and almost 400 thousand acres by the year 2,000.[12] In this respect, the western states face a far different future than the midwestern and eastern coal belts. Outside the West, approximately 66 percent of new coal production will pour from underground mines; in the West, all but a small portion of new coal will be stripped.

Almost all experience with restoration techniques has been in Appalachia and the Midwest where abundant rainfall, vigorously diverse vegetation, and other natural conditions are congenial to vegetative regeneration. In contrast, much of the western coal reserves lies under grassland and brushland where rainfall is typically sparse and vegetation less varied and profuse than among Appalachia's hills or the midwestern plains. In this more fragile biosphere, satisfactory revegetation and other acceptable forms of restoration seem far more problematic. After surveying studies of western restoration in the early 1970s, the House Committee on Interior and Insular Affairs warned that existing restoration technology might not preclude "the possibility of permanently despoiling thousands of acres of productive agricultural lands. . . ."[13] More recently, the NGPRP noted there was still an almost total lack of experience with western restoration and suggested an equally guarded prognosis: "Very little if any land in the Northern Great Plains has been revegetated for sufficient time or with sufficient variety of species to determine potential for success in establishing a permanent ecosystem that will sustain grazing or higher uses."[14] These uncertainties relate specifically to the alluvial valley floors overlying much of western coal and containing lush, highly productive farmlands—premium land to both miners and farmers. Attempts to rehabilitate surface-mined acreage on these alluvial plains have been only partially successful. From the outset of the campaign for national surface-mine regulation, ecologists have insisted that these prime farmlands be given particular protection.

The conclusion commonly reached in technical evaluations of restoration prospects in western land is that the potentials are extremely site specific. That is, each mining site—including alluvial valley ones—involves a slightly different mixture of physical and biological contexts which collectively determine its appropriateness for restoration and, hence, for mining. The "site specificness" of this restoration potential magnifies the importance of administrative arrangements for determining when sites can be mined in provisions of the Strip Mining Control and Reclamation Act; clearly, the constraints and standards imposed upon administrators when making this determination will have a crucial impact upon the environmental consequences of western stripping. This is one of many regulatory issues which illustrate how closely the policy-making structure of regulation affects the environmental outcomes of coal development.

Public Law—A Closer Look

In several respects, the Strip Mining Control and Reclamation Act invites the characteristic problems found in other regulatory programs. These features, fundamental to the bill's conception, need not evolve into major impediments to sound regulation but they are potential structural flaws which facilitate such impediments.

The first of these problems is the *generous discretionary authority customarily granted to federal and state administrators in critical environmental determinations*. The language of the law attempts to hedge this discretion with statutory standards but still delegates enormous discretion to administrators in making decisions about the appropriateness of stripping to specific ecological sites. Among the most prominent of these discretionary provisions are:

● authority to the states to designate land as "unsuitable for stripping" or to withhold such designation according to whether state authorities make a number of judgments regarding the land's physical character and restoration potential.

• authority to the states to permit mining operations west of the 100th meridian on alluvial valley floors providing they find that "mining operations would not interrupt, discontinue or preclude farming . . . nor materially damage the quality or quantity of underground or surface water. . . ."

• authority to permit mining on "prime farmlands" if the regulatory authority "finds in writing that the mine operator has the technological ability to restore the area within a reasonable time to equal or superior form of productivity."

The second characteristic problem is that the *regulation depends upon federal supervision of state regulatory programs and permits states to control mining on federal land subject to federal oversight*. This two-tier regulatory arrangement, a fundamental structural principle, places primary responsibility upon the states to control stripping in their jurisdiction with federal participation confined to technical and financial assistance, administrative oversight, and provision of broad statutory standards for state programs. Specifically:

• States wishing to assume jurisdiction over surface mines must submit programs corresponding to the act's standards for federal approval.

• States assume primary responsibility for administering their own programs, once they are approved.

• Federal authorities will manage regulatory programs only if states fail to submit a satisfactory regulatory plan, request federal management, or "fail to implement and maintain an approved program."

• States may enter into agreements with the federal government which permit the states to regulate stripping on federal land, provided the federal government retains authority to approve or disapprove plans and the power to designate federal lands "unsuitable for coal mining."

Yet another potential structural flaw of the act is that the *state regulatory programs, restoration research, and demonstration projects, together with some costs of restoration itself, are substantially federally funded*. This flow of federal dollars into state restoration activities takes several forms:

• Authorization is provided for the return of up to 50 percent of the fees collected by the federal government for restoration of orphan banks to the states for their expenditure to this end.

• Authorization is provided for $10 million a year in fiscal 1978 through 1980 to "initiate regulatory procedures and administration of the program"; $10 million a year for 15 years for hydrologic studies; $20 million in fiscal 1978 and $30 million each in fiscal 1979 and 1980 for grants to prepare regulatory programs.

• Authorization is provided for $15 million in fiscal 1978, to be increased by $2 million in each fiscal year for six years, for specific research and demonstration projects of "industry-wide application at mining institutes."

These authorizations depend upon subsequent congressional appropriation to become a reality; experience with other federal environmental control programs indicates, however, that such commitments are seldom neglected and will, consequently, be substantially honored.

Looking at these provisions from the perspective of past regulatory experience, and with regard especially for the bias in the traditional formula for coal management, large questions inevitably arise. First, placing responsibility for federal oversight of the program within the Department of the Interior means that the Secretary of the Interior, usually acting through subordinates, will assume much of the discretionary power inherent in the legislation. Interior, long the "department of special favors" to mineral interests, has strong institutional inhibitions against stringent control of mining, even on environmental grounds, and would seem at best an indifferent guardian of the new law. Moreover, the abundant discretion vested in administrative officials when making extremely sensitive environmental decisions would seem to invite pressures that could blunt the impact of the law's environmental safeguards—this in spite of apparently explicit statutory standards to assure the contrary. A case in point is the discretion to determine that a particular mine site is or is not capable of restoration. Although the law bristles with explicit requirements that a wealth of relevant data be assembled and pondered, administrators are still left to determine whether the data conforms to the substantive statutory requirements and—more significantly—whether it is sufficient to support a prohibition on mining. A strong predilection has existed among state officials—who will normally make these determinations—to permit mining and, in making the determination, to resist any other choice unless the most compelling, comprehensive data can be marshaled to the contrary. The House Insular Affairs Committee warns:

> Experience has shown that without a thorough and comprehensive data base presented by the applicant, and absent analysis and review both by the agency and by other affected parties based upon adequate data [sic], this judgement is apt to reflect the economic interest in expanding a State's mining industry.[15]

Thus, discretion can readily permit the utilization of technical data ostensibly conforming to the law yet substantively inadequate and biased toward mine development. Moreover, administrative oversight must be equally vigilant throughout mining operations—and this moves again into discretionary latitudes—because mining company negligence, once permitted, can produce immediate, irreversible ecological damage.

The act, in common with other federal environmental regulations, pins the integrity of the program largely upon the ability of the federal government to exercise effective oversight over state agencies and upon the good faith of state governments to enforce the law operationally. The characteristic political difficulties with federal regulation of state government have been discussed. Besides the considerable ability of the states to resist such regulation, this arrangement may also be deficient in permitting primary enforcement responsibility to rest upon governments with a poor record of enforcing their own past strip-mine regulations. A few states had admirable records. Most, in the words of a congressional investigating committee, were "generally ineffective in bringing about necessary reclamation of the disturbed land areas."[16] The situation described by a coal company official in West Virginia was clearly typical. Asked by the committee whether it "was his impression that, whatever the wording of the law . . . or the way it is administered, the primary criterion is to enable the operator to maximize his profit," the official responded emphatically: "I think that is unquestionable."[17] The act will clearly require most states with existing mine regulatory laws to make their programs substantially more stringent. Consultants to the CEQ indicate that more than half these states will have to upgrade their standards substantially to severely.

In light of past state attitudes toward mine regulation, the act's provisions which permit the federal government "to enter into cooperative agreement with a state for state regulation of strip mining on federal lands" will be a critical matter. Especially in the West, where most coal reserves lie beneath public domain or otherwise rest in federal control, the manner in which these agreements are drawn and enforced will apparently largely determine the character of mining.

Finally, the act's multiple provisions for federal funding of state activities to implement the law establish another dimension of bureaucratic clientelism in federal-state relationships. These arrangements assure that state mine regulatory agencies will be a major constituency of the Department of the Interior, which will in all likelihood exhibit the same solicitude for their viewpoints found when such clientelism develops in other sections of the federal bureaucracy.

The social costs of stripping are not confined to such matters as immediate environmental impacts, regulatory problems, and potentially irreversible ecological deterioration. Stripping also involves less direct, though no less important, costs measured in terms of human and environmental resources diverted from other uses to support mining and its possible secondary activities. These indirect costs are also particularly important to the western states because they dominate public consciousness as much as the immediate environmental risks of stripping and hence are widely regarded as another price of coal development.

Lost Resources: Land and Water

The western public is acutely conscious of its unique natural heritage: vast rolling plains, clear air, broad uncluttered horizons, and a myriad of bountiful yet fragile ecological systems. Much of the nation's remaining unsettled rangeland, together with its abundant game, forests, and meadows, lies directly over rich coal seams. The western coal lands surround mountains and foothills among the loveliest within the continental United States. Grassland, the most common ecological system in the western coal country, abounds in animal and plant life. A description of these grasslands is resonant in images of a national wilderness forever vanished elsewhere:

> In the wetter North Dakota area, unplowed native grass is dense and ranges knee high to waist high. In the drier areas of southeastern Montana and northeastern Wyoming, it is little more than ankle deep and mostly blue grass, western wheat, and needle and thread grasses. . . . Throughout the grassland there are also forbs (herbs) that mix the bright colors of their blossoms with the various shades of green. . . . There is big sagebrush for which the West is famous. Wildlife includes antelope, mule deer, jackrabbits, prairie dogs, black-footed ferrets, meadowlarks, sharp-tailed grouse, and birds of prey. Around cropland are ring-necked pheasant and red fox. The land and pothole country of northwestern North Dakota and northeastern Montana is part of the North American "duck factory." The endangered whooping crane passes through here annually.[18]

While grasslands are most vulnerable to stripping, other western ecological systems may also yield to the miner's augers and shovels. Table 3.2 is an estimate of the various western ecological systems upon which strip-mining is practical. It is impossible to predict confidently how much of these ecosystems might be successfully restored or whether successful restoration would substantially reproduce the land's original character. Many westerners fear that mining, no matter how carefully regulated, will destroy forever the aesthetic qualities which permeate the western plains and endow them with their particular ambience.

Water, traditionally dear as gold in the semi-arid western ecosystem, must also be diverted in large quantities to support a new western coal economy. Surface mines rarely require large volumes of water but the slurry systems that may be built to move the coal eastward and the coal conversion plants which may appear near the mines could draw huge draughts of western water. A moderately large coal slurry system requires approximately 600 to 800 acre-feet of water to transport a million tons of coal. Estimates indicate that the water required to transport by slurry half the coal exported from the western Great Plains under maximum production conditions would be about 186,000 acre-feet a year; this water demand, considered plausible in light of the many incen-

TABLE 3.2

Proportion of Major Ecosystems Capable of Surface Mining in the Northern Great Plains

Ecosystem Category	Total Acres in Category (1,000 Acres)	Surface- Minable Acres in Category	Percent of Surface Minable (Acres)
Grasslands	61,646.8	1,515,725	58.3
Flood plains	3,755.6	54,597	2.1
Badlands	3,022.8	31,198	1.2
Brushlands	15,572.0	725,363	27.9
Ponderosa pine	3,664.0	272,986	10.5
Mountains	3,938.8	0	0.0
Totals	91,600.0	2,499,869	

Source: Northern Great Plains Resources Program, *Effects of Coal Development in the Northern Great Plains*. Report prepared for the U.S. Department of Agriculture, April 1975. NTIS Document No. PB-269 863.

tives for future slurry construction, would drive up the cost of water for agriculture and would directly compete with farming.[19] Conversion of coal to electricity or synthetic gas, in addition, depends critically on the availability of water. Extensive electric generation or coal gasification, consequently, would also compete with agricultural water demand. Given limitations on the reserves of surface and underground water in the western coal areas, it appears almost inevitable that extensive slurry and coal conversion operations would force a tradeoff with future agricultural production. It is this prospective competition between coal production and agriculture that incites some of the most vigorous opposition to stripping from the western agricultural community.

Strip mining and its associated activities have stirred the greatest public attention among all the environmental aspects of coal development— understandably, for the ecological risks are vividly apparent. The coal development will also affect the nation's air quality, however, and thereby influence public health. Moreover, the nation has already invested billions of dollars in protecting air quality and has made commitments to air pollution standards that are unlikely to prevail if massive new coal utilization occurs. Thus, the sunk costs in air quality, as well as existing air quality standards, have to be counted as probable losses if a national coal utilization program unfolds along projected lines.

CLEAN AIR AND DIRTY COAL

Coal is the dirtiest of all fossil fuels when burned by existing methods. Used in industrial boilers and electric generating plants, it pours large volumes of carbon monoxide, sulfur oxides, hydrocarbons, nitrogen oxides, and particulates into the atmosphere, together with lesser amounts of radioactive material, heavy metals, and trace elements. The nation's air pollution program, embodied in the Clean Air Amendments (1970), contains air quality standards intended to limit the emissions from coal burning facilities because these stack gases, when uncontrolled, constitute a potential health hazard and, under conditions by no means rare, can be lethal.

The widespread impact upon air quality of coal conversion among major coal users is suggested in Table 3.3, which deals with electric generating plants, the principal national source of air pollutants from coal. This table indicates by state and region the number of electric generating plants presently under FEA orders to convert to coal, those under consideration for such orders, and planned new facilities (virtually all of which will be coal burning). In addition, approximately 1,200 industrial boilers are eligible for coal conversion; most of these are in the Northeast, the Middle Atlantic, and the upper Midwest.[20]

The Sulfur Oxide Dilemma

It appears to be virtually impossible to increase coal consumption to levels desired by most energy planners without lowering existing state and national air quality standards for sulfur oxides, the most common and immediately dangerous of air contaminants associated with coal combustion. Conversely, a decision to enforce existing air quality standards will, in effect, be a decision to constrain coal production. The reciprocity between coal and air pollution policy is such that no short-term technological fix in pollution control technology can simultaneously eliminate the undesirable consequences of both situations.

The incompatibility between significant new coal utilization and existing air quality standards arises from several circumstances. In many national air quality regions, particularly in the Northwest, Southwest, and Middle West, air quality is often "undegraded"—that is, the highest possible, and significantly above the minimum acceptable standards set by Washington or the relevant states. Coal burning in existing or planned installations in these regions will cause a significant deterioration of air quality, not necessarily below the minimum acceptable existing standards but far below present air conditions. In these situations, federal and state officials would have to permit modest to very significant air degradation—an event which is presently prohibited by many states and, in some cases, by Washington. In other geographic areas, primarily heavily industrialized and urbanized regions, air quality does not now meet existing state or federal

TABLE 3.3

Prospective New Coal-Fired Energy Systems, by State and Region, 1977–85

Region and State	Electric Generating Units		
	New	Under FEA Orders	Under FEA Study
Northeast			
Maine	—	—	—
New Hampshire	—	—	2
Vermont	—	—	—
Massachusetts	—	—	8
Rhode Island	—	—	—
Connecticut	—	—	8
Middle Atlantic			
New York	3	—	9
New Jersey	—	—	5
Pennsylvania	4	—	2
Delaware	1	—	4
Maryland	1	—	8
District of Columbia	—	—	—
South Atlantic			
Virginia	—	—	10
West Virginia	5	—	—
North Carolina	3	2	1
South Carolina	3	—	—
Georgia	5	—	5
Florida	6	—	2
South Central			
Kentucky	16	—	—
Tennessee	—	—	—
Alabama	8	1	—
Mississippi	5	—	—
Arkansas	5	—	—
Louisiana	8	—	—
Oklahoma	12	—	1
Texas	36	—	—
North Central			
Ohio	8	—	—
Indiana	10	—	—
Illinois	7	—	4
Michigan	15	—	1
Wisconsin	8	1	—
Minnesota	6	—	1
Iowa	5	4	5
Missouri	8	—	5

TABLE 3.3 (Continued)

Region and State	Electric Generating Units		
	New	Under FEA Orders	Under FEA Study
Nebraska	4	1	2
Kansas	7	2	11
Rocky Mountain			
Arizona	10	—	—
Montana	2	—	—
Wyoming	8	—	—
Idaho	—	—	—
Colorado	10	—	—
No. Dakota	6	—	—
So. Dakota	—	—	—
Utah	8	—	—
Nevada	4	—	—
New Mexico	2	—	—
Pacific			
Washington	—	—	—
Oregon	1	—	—
California	—	—	—
Alaska	—	—	—
Hawaii	—	—	—
Total	250	11	94

Source: National Coal Association, "Achieving the President's Goal for Increased Coal Production and Use," mimeographed, May 12, 1977.

standards and consequently further air deterioration is not permitted and any new pollution sources must be counterbalanced by a decrease in pollution from an existing source. It will be very difficult, and often impossible, for new coal combustion to be introduced in these areas without further air quality deterioration because the reduction in existing air pollution sources intended to balance the new pollution load will not be possible within a reasonable time.

Many states, moreover, have adopted air quality standards for sulfur oxides and other pollutants more stringent than federal ones; in these areas, new coal combustion will be impossible unless the states yield to lower federal standards—a situation many environmentalists, and others in the affected areas, oppose.

What imparts a particular intensity to the argument over the environmental impact of new coal utilization is the apprehension among many observers that lowered air quality standards will be, even more than an ideological defeat for the environmental ethic, a direct menace to public health. While proponents of stringent air quality standards have sometimes indiscriminately asserted that any reduction of existing air quality will produce a health hazard, there is good reason to believe that in some heavily populated urban areas such a threat is plausible. Commenting on the coal conversion goals of the NEP, for instance, the OTA observes:

> Some areas such as Southern California will not be able to burn coal without creating serious health hazards. . . . Most Southern California powerplants are in densely populated air basins where air quality is already bad. Conventional coal plants, even with the best available control technology, are likely to emit more particulates and sulfur than State law allows.[21]

Reference to "best available control technology" in the OTA's assessment is particularly significant. Control technologies frequently are invoked by proponents of new coal utilization, often with bland disregard for the difficulties involved, as the technical fix to pluck the nation from the ecological thicket created by coal utilization. A confidence that technology can cure its own excesses is deeply ingrained in American culture and appeals to national pride in American know-how. Nevertheless, the ability of existing control technologies to prevent a significant sacrifice of air quality standards to coal consumption is extremely uncertain.

The Limits of Control

President Carter's NEP, like most coal development proposals currently advanced, proposed to offset the environmental impacts of new coal utilization largely by requiring "the installation of the best available control technology in all new coal fired plants" and to encourage its utilization in existing installations converting from other fuels. (New control technologies would for many reasons be almost mandatory for most converting installations in any case.) The best available technology is currently considered to be flue-gas scrubbers to purge stack emissions of most sulfur oxides and electrostatic scrubbers to control particulates. Industrial and utility interests have vigorously opposed flue-gas scrubbers because they are assertedly unreliable and unreasonably expensive. National experience with flue-gas scrubbing is so limited and technical studies are still so tentative and conflicting that it is impossible to make a convincing case for their reliability or economic acceptability; recently, the OTA and the GAO have expressed serious reservations about both their inherent effectiveness and the likelihood that such scrubbers, even if they perform satisfactorily, will be available sufficiently soon to serve as a near-term control on stack emissions.[22]

Two other undisputed environmental problems are associated with scrubbers. First, this technology does not control (even in combination with electrostatic precipitators) such potentially dangerous emissions as radioactive materials, hydrocarbons, carbon monoxide, and many heavy metals in a gaseous, liquid, or solid state. Thus, concludes the GAO, "if coal production is increased significantly, further environmental degradation will take place despite the strong pollution control measures proposed in the plan because many pollutants emitted from coal burning are not regulated and cannot be controlled even using the best available control technology."[23] Moreover, the extensive use of scrubbers would result in the massive production of toxic sludge; the GAO has estimated, for example, that the NEP with scrubbers would produce 230 million tons of this sludge in 1985.[24] Disposal of such sludge is technically possible and in limited quantities has already been achieved. Little governmental planning, however, has been invested in a problem of this magnitude nor, indeed, has this specific problem been generally acknowledged by the current administration.

Carbon Dioxide and the "Greenhouse Effect"

In addition to the familiar environmental problems in coal utilization, another related one has recently surfaced. Environmental scientists have long pondered the potential long-range effects upon the earth's climate of the continually mounting volume of carbon dioxide pouring into the atmosphere from growing world fossil fuel consumption. One frequently suggested consequence is the "greenhouse effect." Some experts have proposed that increasing carbon dioxide may over many decades prevent progressively more heat from radiating back into space from the earth's atmosphere and thereby cause a gradual ecologically critical increase in the earth's average temperature. In mid-1977, William D. Rordhaus of the CEQ concluded that the trend toward heavier use of fossil fuels might double the level of carbon dioxide in the atmosphere early in the next century and result in a multitude of undesirable environmental consequences; at about the same time, several scientific spokesmen for the U.S. Energy Research and Development Administration (ERDA) were advancing the same argument before Congress. The most substantial support for these theories, however, was offered by a special panel of the National Academy of Sciences in late 1977. The panel warned of the possibly grave consequences of increased coal burning in the United States. Estimating that by the year 2,000 the carbon dioxide content of the air will have risen 25 percent above levels prior to the Industrial Revolution, the panel asserted that the potentially ominous consequences of this carbon dioxide concentration were sufficient to raise serious reservations about the desirability of a new national coal utilization drive.[25]

At the moment, the greenhouse effect is little more than intelligent speculation based upon tentative, and still controversial, data. The breadth and vigor of scientific concern for the issue, however, justifies its inclusion among the compendium of potential environmental risks to further new coal utilization.

The East-West Impacts

The regional distribution of environmental impacts from coal utilization will also be affected by air pollution policy. The extent of western coal utilization will in some measure be sensitive to future federal environmental decisions. If Washington continues its present sulfur oxide policy in respect to both air quality standards and acceptable control technologies, pressure to develop extensively the western coal seams, with their generally low sulfur content, would almost assure a pervasive, short-term increase in coal mining throughout the western reserves. If, in contrast, federal authorities were to relax air quality standards for sulfur oxides, it is quite possible that greater amounts of high-sulfur eastern coal could be consumed with a resulting decrease (though not an elimination) of pressure on western sources. A similar regional difference in impacts would be felt by changes in the EPA's current regulations for control technology on sulfur oxides. Western coal is now desirable because the EPA exempts those installations burning low-sulfur coal from requirements for scrubbers. Should the EPA require that *all* industrial and utility boilers utilize scrubbers, however, regardless of the coal used, this would lessen the attractions of western coal considerably because its economic advantage to users would be considerably lowered.

The extent to which regional demands for new coal are sensitive to federal air pollution policy is but one additional illustration of the intricate, and unavoidable, nexus between the decisions of public agencies and the environmental consequences of coal utilization.

SUMMARY: THE CLOUDED ENVIRONMENTAL FUTURE

The conclusion is inescapable that rising national coal consumption will create major adverse environmental impacts that can only partially be ameliorated by regulatory programs. Some of these ecological ills would occur as a result of rising coal consumption normally to be expected even without a federal coal conversion program. A vigorous federal effort to increase domestic coal utilization further would intensify the environmental impacts proportionally. The role of federal and state governments will be crucial, not in preventing environmental damage from coal utilization, but in containing the damage.

The policy implications of this situation are obvious—in spite of the desire of public officials to obscure the harsh implications. First, in the words of the OTA, coal development will mean a "deliberate choice between increased use of coal and air-quality goals . . . in the short run."[26] Second, the degree to which the ecological damage of coal utilization can be contained depends partially on unproven technologies but, equally importantly, also upon a regulatory process and a substantive regulatory program (particularly for strip mining) with numerous inherent potentials for failure. Third, the environmental burden of coal production will fall with particular severity upon the western states. Finally, a concern for the environmental impact of coal production constitutes the most serious impediment to massive new domestic coal utilization. To the extent that the ecological constraints upon coal utilization cease to attract a vigorous, articulate and politically skilled constituency, the way to unrestrained coal utilization will be made that much smoother.

Ecologists fear that the energy crisis, coupled with public awareness of the nation's vast coal reserves and its reluctance to sacrifice its enegy-consuming conveniences, will undo the gains in national environmental policy made during the early 1970s. In this there is at least some justifiable apprehension. The energy crisis has provided proponents of massive new fossil fuel development with a potentially powerful symbol with which to attack ecological programs otherwise well rooted in public esteem. It is entirely plausible, though by no means predictable, that coal will become the first energy domain in which the energy crisis will create sufficient political leverage to permit opponents of existing air and water quality programs to roll them back. Indeed, this could be done even as new regulatory programs roll from Congress to prevent the environmental ravages always inherent in uncontrolled coal utilization.

NOTES

1. NEP, p. 67.

2. The logic of the "standards and enforcement" approach is summarized in Walter A. Rosenbaum, *The Politics of Environmental Concern*, 2nd ed. (New York: Praeger Publishers, 1977), chapter 5.

3. Breyer and MacAvoy, op. cit., p. 15.

4. Ibid., p. 119.

5. Wilson and Rachal, op. cit., p. 10.

6. Schultze, op. cit., p. 56.

7. Wilson, op. cit., pp. 77-103.

8. A very thorough examination of strip mining's impact is found in U.S., Congress, House, Committee on Interior and Insular Affairs, *Surface Mining Control and Reclamation Act of 1976: A Report* (Washington, D.C.: Government Printing Office, 1976), H. Rept. 94-1445.

9. Production figures for strip-mine coal may be found in U. S., Department of the Interior, Bureau of Mines, *Minerals Yearbook*.

10. ICF Incorporated, *Final Report: Energy and Economic Impacts of H.R. 13950* (Washington, D.C.: ICF Incorporated). Prepared for the Council of Environmental Quality. Provisions of H.R. 13950 are not significantly different from Pub. L.95-87 in respect to reclamation requirements.

11. House, H. Rept. 94-1445, op. cit., pp. 29-30.

12. Northern Great Plains Resources Program, op. cit., p. 56

13. House, H. Rept. 94-1445, op. cit., p. 58.

14. Northern Great Plains Resources Program, op. cit., p. 53.

15. House, H. Rept. 94-1445, op. cit., p. 43.

16. Ibid., p. 12.

17. Ibid., p. 21.

18. Northern Great Plains Resources Program, op. cit., pp. 48–49.

19. Ibid., p. 75

20. Information provided to author by U. S., Department of Energy, Freedom of Information Office.

21. OTA, p. 184.

22. See U.S., Comptroller General, *An Evaluation of the National Energy Plan* (Washington, D.C.: Government Printing Office, 1977), chapter 5; and Congress, Office of Technology Assessment, op. cit., pp. 157–59.

23. Comptroller General, *An Evaluation of the National Energy Plan*, op. cit., p. 5.33.

24. Ibid., p. 5.17.

25. *New York Times*, June 3, 1977.

26. OTA, p. 157.

4
Coal and Foreign Policy

Coal issues do not end at the water's edge. Domestic coal production is, and will continue to be, highly responsive to U.S. diplomatic and military policy. Decisions concerning the extent to which the United States can and should be independent of overseas energy resources, for instance, will directly affect domestic coal production. Moreover, domestic energy technologies are interdependent; policy affecting one energy sector affects all. This is particularly true of the relationship between nuclear energy and coal. Commitments now being made in Washington concerning the character of desirable future nuclear technologies to be domestically developed and internationally promoted will impact upon coal development over many decades.

The competing foreign and military policy doctrines whose resolution directly affects coal production represent, in good part, conflicting definitions of international reality and divergent priorities between domestic and foreign needs. Since World War II, foreign and military issues have customarily subordinated domestic problems in White House and congressional policy agendas. If this is still the case, then U.S. strategic doctrine in diplomatic and military affairs can be expected to exert substantial influence on coal development even though much of the domestic debate over coal utilization focuses almost exclusively upon its domestic implications—environmental impacts, for example. A resurgence of domestic priorities in Washington's policy agenda, however, would probably force a very different energy perspective in which the military and diplomatic aspects of coal development would be given lesser weight and the domestic impacts and constituencies associated with coal would be accorded more attention.

COAL AND NATIONAL SECURITY

The Nixon presidency has been interred but the ghost of President Nixon's "Project Independence" continues to haunt current Washington energy discussions. This is particularly true of the coal issue. The president's proposal of national energy independence within a few decades has proven a chimera quite properly put to rest. The various scenarios proposed in the project to achieve this independence by 1985—or any date within a century—have proven indefensible under close scientific, economic, and technical scrutiny. Even the energy lobby, once the most vigorous promoter of this independence, no longer pursues the illusion.

But two salient ideas in Project Independence continue to color discussions of coal policy. These ideas, among the principal suppositions embraced by those who can be called the "hard liners" on energy policy, are, first, that national security—military and economic—should be the primary consideration in a domestic energy policy, and, second, that coal must be a major resource in obtaining this future security. Almost from the time Americans were forced to recognize the "energy crisis" during the oil embargo of 1973-74, these viewpoints have been advanced vigorously, by Presidents Nixon and Ford, by the Pentagon and State Department, by ERDA and, usually, by spokesmen for the energy lobby. President Carter, less wholly preoccupied than his immediate predecessors with the military implications of the energy crisis, nonetheless demonstrated how deeply a consciousness of the international effects of the energy situation affected his thinking when he listed as the first overriding objective in his NEP the reduction of "dependence on foreign oil and vulnerability to supply interruptions."[1] In its most extreme expression, this hard line psychology has led to calls for immediate, vast expansion of all domestic energy resources, including a rapid development of new production technologies (like breeder reactors) and commercial perfection of experimental ones (like coal liquefaction and gasification).

The opposition to this hard line comes from those, including especially environmentalists, who do not accept this definition of national energy priorities.[2] The collision of values is quite direct. The Environmental Agenda Task Force, recently sponsored by the Rockefeller Brothers Fund, spoke for the nation's major environmental groups in taking specific exception to this hard line viewpoint: "The environment must be recognized as a top national priority on a level with defense, employment, health, education, and commerce."[3] The implications of this value conflict are significant for future national energy policy. Weighing the environmental consequences of coal and nuclear development equally with their international impacts would inhibit a "crash" program of domestic coal development and the rapid production of new nuclear generating facilities. At the same time, greater environmental concern would encourage a more cautious approach to the substitution of coal for nuclear power lost through failure of the domestic nuclear power industry to produce its anticipated share of future electric power demand.

These differing perspectives can be illuminated by examining how specific foreign and military policy issues relate to coal. It will be apparent that differing definitions of international and national priorities often dictate concepts of desirable domestic energy development, including coal policy specifically. It is also evident that coal, like other U.S. energy resources, is so deeply involved in international policy that it can never be viewed entirely as a domestic problem.

"Our Insurance Policy"

The administrations of Presidents Nixon, Ford, and Carter, however much they may have otherwise disagreed, spoke with a common voice on the stretegic importance of energy development, often with special attention to coal. The dominant theme, echoed especially by major administration spokesmen in energy and military affairs, was early sounded by Frank Zarb, head of the FEA in the latter Nixon years: "Coal is the United States' insurance policy. Coal is there, under our control."[4] Under the Ford Administration, coal's environmental risks seemed far less impressive to the White House than its strategic attractions. "It is far easier to reclaim the environment than it would be to reclaim our independence from foreign oil sources a decade from now," explained Secretary of the Interior Walter Kleppe.[5] More recently, the Pentagon has taken the position that domestic energy development in general must be considered a critical strategic issue. In the words of Secretary of Defense Harold Brown, "lagging fuel supplies pose the single largest threat to national security."[6] The association of coal development with national security is so deeply rooted in current Washington policy discussion that it has become in effect a first premise from which all subsequent analysis proceeds.

There has been, in fact, virtually no vigorous challenge to the argument that increased coal production would, in theory, increase U.S. strategic security to some extent by making the national less vulnerable to foreign energy blockades or price escalations—not even a challenge from environmentalists. Rather, the argument turns on *how long* and in *what technological form* coal development should continue. Upon these matters there is bitter dispute.

The Merits of "Redundancy"

The U.S. business community has been an enthusiastic proponent of broad and rapid domestic energy development, for its presumed strategic value and because it would allegedly buffer the United States from the shocks of repeated, short-term escalations in crude petroleum prices. The domestic economic implications of these chronic price instabilities have been widely explored; most economists agree that greater predictability in crude petroleum prices and a diminished reliance upon imported crude oil, with its large impact on the U.S. balance of trade, would be desirable. Of greater importance to domestic coal policy is the broad pressure exerted across the U.S. business community to simultaneously develop as many domestic energy sources as soon as possible, and to the limits of U.S. ability. The pull-out-all-the-stops approach, which identifies U.S. strategic and military interest with immediate massive energy production, is defended by the Committee on Economic Development (CED) as "prudent":

> There is much merit in the argument for redundancy in the face of uncertain energy supplies . . . a prudent energy policy should seek insurance against misfortune. Thus, except for the constraint of scarce manpower or material resources, there is no need to choose between coal or nuclear power or between oil shale and gasification, at least until their relative costs and merits are clearly understood. *We believe that doing too much to achieve energy independence is a more acceptable risk than doing too little. . . .*[7]

Such an argument for energy "redundancy," issuing from an organization identified with the more liberal elements among U.S. big business, suggests the breadth of its support within the U.S. business community.

Not surprisingly, in the business sector the most articulate proponents of rapid expansion of domestic energy resources for strategic reasons are the petroleum and electric power companies. Over the last decades the coal industry—traditionally highly competitive and even anachronistic in its large number of small entrepreneurs—has become more highly concentrated; the preponderance of currently mined coal is produced, with few exceptions, by subsidiaries of large conglomerates. Twenty companies produce more than half the U.S. domestic coal output; among these major producers are corporations owned by Continental Oil, Exxon Corporation, Occidental Petroleum, and Gulf Oil. Gulf and Exxon, especially, have vigorously promoted new energy production through extensive advertising in major national newspapers and news magazines. Joining the oil companies in this promotion have been such major utility spokesmen as the Edison Institute, representing the large investor-owned companies, and regional associations including the American Electric Power System, a consortium of midwestern utilities (which admonished readers that "America has more coal than the Middle East has oil. Let's dig it!").

This energy lobby's intense political pressure and media campaigns to expand quickly domestic energy resources have been opposed by environmentalists, conservationists, and opponents of nuclear power generating plants. It is the push for the simultaneously massive development of both coal and nuclear power that acutely alarms the critics of this strategic doctrine. Part of this apprehension is grounded in the special menace which many Americans impute to nuclear generating plants. But the opposition also reflects a recognition, shared by many experts outside environmentalist circles, that the nation can make choices and tradeoffs between nuclear and coal power that could preclude the simultaneously massive development of both. Moreover, the strategic national interest might well be served by short-term and controlled development of one, or both, fuel sources while new technologies are developed. In short, the policy options are more complex than might be implied by a straightforward equating of U.S. strategic interest with pervasive growth of all domestic energy producing sectors.

The options available in enlarging U.S. domestic energy production illuminate the interdependence of coal and nuclear fuel policies. This illustrates as well the sensitivity of these   domestic   fuel sources to foreign and military policy commitments.

THE COAL-NUCLEAR NEXUS

The only domestic energy sources which can be substantially enlarged before the turn of the century and coal and nuclear energy. Coal, and possibly nuclear energy, might eventually play a significant role as a primary fuel for industrial use but the greatest importance of both will be to generate the greatly expanded electric power expected through the year 2,000. A fact less widely publicized by proponents of expanded domestic energy production is that coal and nuclear power are, and might continue to be, economically competitive in the domestic market. There have been widespread predictions that nuclear power, not coal, will be the primary electric generating fuel of the future. Consultants to the Senate Committee on Interior and Insular Affairs recently suggested that nuclear energy may be "the key factor displacing oil in electric power generation" and, thus, a replacement for anticipated new utility coal demand.[8] A primary justification for the U.S. development of the controversial breeder reactor has been its presumed advantages, both economic and technical, over coal in power generation. Presently, coal and nuclear energy are often economically competitive and, in the South and Northeast where local coal supplies are very meagre, nuclear power is economically more attractive.

The nuclear power industry, nonetheless, is in trouble—economically, environmentally, and technically. Its future development is clouded with enormous uncertainties whose resolution will largely depend upon U.S. foreign and domestic policy decisions. These, in turn, will impact upon coal production by making domestic coal utilization more, or less, economically and technically attractive when compared to nuclear power. Generally, the tradeoff between nuclear power and coal as domestic electric generating fuels is rather direct: the more one increases in utilization, the less the other is likely to be utilized. For environmentalists, this tradeoff constitutes a Hobson's choice many seem not to recognize; others deny its existence by asserting that other energy strategies can obviate such an unpalatable option.

A Sick Industry

"The midterm future of [nuclear power] is in . . . doubt," notes the OTA. "Rising costs, licensing delays, and slippage in construction schedules have caused the nuclear industry to place a de facto moratorium on orders for new plants after 1985. . . ."[9] This moratorium on new orders, if continued, would mean the end of new plant construction and in broader perspective would cause a major revision of projected U.S. energy needs and resources. In California, for example, the state's governmental energy planning had been predicated upon the construction of twelve new "nukes" now apparently cancelled or relegated to uncertainty; federal energy planning, as we shall shortly observe, was similarly premised upon the appearance of new nukes now highly questionable. Significantly, utility companies once planning to bring nuclear generating installations "on line" in the future are now turning to coal instead as their future energy source.

The reasons for the nuclear power industry's sagging prospects are manifold. Economically, nukes are far more costly and less productive than was once supposed. Construction costs have climbed steeply within a decade; in 1972, a nuke could be constructed for about $300 per kilowatt of capacity but in 1985 (the time when plants now started would finally be finished under optimal circumstances) the cost is anticipated to be $1,135 per kilowatt of capacity.[10] Existing nuclear plants usually operate at little more than half of planned capacity—the national average is 57.5 percent. Complexities in siting and licensing plants, difficulties in obtaining the necessary safety and environmental clearances, and additional delays caused by litigation initiated by opponents of nuclear power have stretched building time for the average nuclear plant to between 10 and 12 years. In comparison, new coal-fired electric power installations can be constructed in about six years and, in most regions of the United States, produce power 20 percent cheaper per unit.[11] Facing these dismal prospects, it is understandable that 145 of the 170 domestic nuclear units once projected have been deferred for several years or cancelled indefinitely. In 1977 only three new reactors had reached operational state and only 10 more were under construction.

Despite these besetting ills, nuclear power is still cost-competitive with coal, even superior, in several major U.S. markets. Moreover, it is quite conceivable that governmental policies relieving the industry of the long procedural delays in plant construction, encouraging the development of new, economically attractive technologies, and increasing the demand for nuclear technology abroad could improve the industry's economic prospects dramatically. Comparing the future economic prospects of nuclear and coal energy, a recent Ford Foundation study concluded that the relative advantages of one fuel over another in the domestic market could easily tip one way or another according to governmental domestic or foreign policies:

> . . . despite large uncertainties, nuclear power will on the average probably be somewhat less costly than coal-generated power in most of the United States, or, more precisely, in areas that contain most of the country's population. . . . The cost factors are sufficiently uncertain, however, that the balance could shift or increase or eliminate the small advantage that nuclear power appears to have on the average in present estimates.[12]

The sensitivity of coal and nuclear fuel prices to alterations in current governmental policies means among other things that all projections of future U.S. energy supply and demand incorporating estimates of these fuel availabilities and markets are highly provisional and extremely liable to changes introduced by governmental design rather than market forces alone. This makes the current public policies affecting nuclear fuel, especially, relevant to future coal planning.

Current Governmental Policy

Presidents Nixon, Ford, and Carter have promoted nuclear energy, both explicitly and implicitly, as a major factor in meeting long-term U.S. energy demand and thereby helping to improve the nation's international security. Indeed, all current federal energy planning is premised upon the existence of growth of a vigorous nuclear power sector. Put somewhat differently, present national coal utilization planning at the federal level assumes vigorous promotion of nuclear power. In the absence of this nuclear development, presently projected needs for coal in the short term as well as the long term would have to be revised sharply upward or conservation would have to be practiced on a greatly increased scale to replace the lost nuclear energy.

The commitment to nuclear power was quite explicit and unambiguous under Presidents Nixon and Ford. Following the guidelines first presented in the Nixon "Project Independence," President Ford called for the completion of 135 new nuclear plants by 1985—a schedule requiring the (impossible) construction of one new plant a month! The Ford administration, recognizing the necessity of federal subsidization of nuclear power to meet an ambitious short-term increase in nuclear power production, proposed a $100 billion Energy Independence Authority that would, among other objectives, achieve this goal. The breeder reactor program, focused on the development of the first U.S. prototype at the Clinch River project, has been the largest single item in federal energy expenditures from the Nixon through the Carter Administrations; by mid-1976 more than $3 billion had already been spent on developing the breeder reactor. ERDA, now absorbed within the new U.S. Department of Energy, has been among the most active bureaucratic proponents of a major U.S. nuclear development program. The Department of Energy, ostensibly too new to have established a record of its own on energy development policy, is widely assumed to be highly favorable to rapid domestic nuclear development.

President Carter's strong, continuing opposition to U.S. development of breeder reactors and other advanced nuclear technologies utilizing plutonium has been widely, and mistakenly, interpreted as an indication that the Carter administration would generally place less emphasis upon nuclear technology as a solution to domestic energy shortages. The president, on the contrary, has been a continual advocate of expanded domestic nuclear power generation that does not utilize the technologies which he believes will encourage international nuclear proliferation. The extent of this commitment to domestic nuclear energy, primarily in the form of advanced light-water reactors, was evident in the NEP. Noting that the United States "must continue to count on nuclear power to meet a share of its energy deficit," the president placed his administration solidly behind new light-water reactor technologies:

> . . . because there is no practical alternative, the United States will need to use more light-water reactors to help meet its energy needs. The Government will give increased attention to light-water reactor safety, licensing, and waste management so that nuclear power can be used to help meet the U.S. energy deficit with increased safety.[13]

The great reliance placed by the Administration on new nuclear facilities coming on line *even without any additional governmental incentives* is suggested by Table 4.1, which indicates present and projected domestic energy supplies; the estimated nuclear contribution in the future does not differ substantially from other current projections though all may be too optimistic in light of the nuclear

TABLE 4.1

Projected Growth in Nuclear-Generated Electric Power Demand under the National Energy Plan, 1976–85: Fuel Balance by Sector (Millions-of-barrels-of-oil equivalents per day)

Energy Source	1976	1985 Without Plan	1985 With Plan
Oil	1.6	2.0	1.3
Natural gas	1.5	.9	.5
Coal	4.9	8.2	8.3
Nuclear	1.0	3.6	3.8
Other	1.5	1.6	1.6
Total	10.5	16.3	15.5

Source: U.S., Executive Office of the President, Energy Policy and Planning, *The National Energy Plan* (Washington, D.C.: GPO, April 1977), p. 95.

industry's present tribulations. According to these estimates, nuclear power output would roughly triple by 1985.

The United States cannot expect a substantial increase in nuclear power generation within the next several decades unless a number of reforms are achieved in procedures for nuclear plant licensing and siting. Long-term nuclear power growth may moreover depend upon development of plutonium-based advanced technologies. The nuclear power industry's future is so dependent upon federal government decisions that the industry currently regards the political arena as the forum where the industry's survival is at stake. Nuclear power issues also arouse, among environmentalists and many others, an especially emotional reaction. Environmentalists and other public interest organizations have bitterly opposed many if not most reforms intended to revive the nuclear power industry; many want the industry's eradication. The resolution of this conflict clearly involves more than the fate of nuclear power; future coal demand and, indeed, future planning for all energy resources will depend upon governmental choices about the nuclear future.

The Current Issues

The Carter administration's proposals for substantive and procedural reform in federal law relating to nuclear plants are regarded by the nuclear industry as little more than a survival package; the industry believes its future prospects will ultimately depend upon development of advanced plutonium technologies. The Carter proposals are directed against the most common obstacles to completing existing planned plants.[14] These reforms include:

- increased safety and security inspection of light-water nuclear plants.
- development by the Nuclear Regulatory Commission (NRC) of firm, predictable siting criteria to preclude mis-siting in the future.
- reform of the licensing process to speed federal, state, and local approval of plants, especially through the use and approval of standard plant designs.
- new labor-management agreements to speed construction time by ending strikes and other labor delays.
- new waste management programs with greater attention to long-term safety and disposal of nuclear wastes.

Through these and other measures relating to light-water plants, the administration believes it can compress the construction time of currently planned nukes to approximately six years, or about half the current actual average. This abbreviated schedule would enhance the economic attractions of nukes. Construction capital would allegedly be more available and generating costs lower;

consequently, nuclear power would enjoy widespread competitive equality with, or perhaps advantages over, coal-generated power.

Proponents of nuclear power regard these measures as little more than a "Band-Aid" compared with the reforms they believe are imperative. The programs they believe will launch the industry securely into the future and guarantee to the nation a dependable, large nuclear power reserve are those which directly conflict with the administration's current foreign policy commitments on nuclear energy. Especially, partisans of nuclear power propose a national commitment to the following goals:

• development of a large domestic technology for recycling and reprocessing plutonium to prevent the depletion of U.S. nuclear energy reserves through continued mining of limited uranium ore.
• rapid creation of a breeder reaction technology.
• creation of a safe nuclear waste disposal strategy.
• expansion of U.S. nuclear enrichment capacity.
• commitment to exporting plutonium fuels, plutonium-based technologies, and the necessary technical assistance to install them overseas.

Environmentalists have been the vanguard of the opposition to governmental reform of light-water reactor policy or new governmental stimulation of plutonium technologies. The former they generally regard as a reckless gamble with public and environmental safety; the latter they consider intolerable. By rallying public opposition to rapid nuclear power development, the environmental movement has been the principal political opposition to the nuclear power industry. It may also succeed in posing for the nation a very difficult choice between nuclear and coal-generated power in which massive environmental damage will be risked regardless of the decision.

The Environmental Opposition

Opinion polls seldom suggest a firm aversion to nuclear power among the American public. In 1976 ballot propositions intended to delay plant development in seven states all failed at the polls; California's Proposition 15, generating the most widely publicized and bitterly fought of these ballot battles, was solidly defeated in what was nationally considered the most significant vote. Nonetheless, environmentalists allied with other public interest groups have so effectively used a variety of political tactics, particularly litigation, to delay plant construction and to arouse public doubt about plant safety that the OTA has warned that "nuclear power may not be a viable source of energy if public acceptance continues to erode."[15]

The environmentalist campaign against expanding nuclear generating capacity proceeds at two levels. At one level, it attacks the nuclear plant's safety and reliability; it warns of dangers and demands additional precautions arising from well-known, and often widely acknowledged, problems within the industry. At another level, it reaches the pitch of a moral crusade in which nuclear power is made to symbolize a dark and menacingly imminent social philosophy threatening to corrupt American political institutions. On this second level, the battle attains the fervor of ideological war in which political philosophies, not technological engineering, are joined and compromise seems to smack of betrayal.

Environmental difficulties afflicting current nuclear generating procedures are public knowledge. The industry has been plagued with unexpected safety problems; legitimate doubt has been created that the reactors are as wholly safe as the industry propagandists allege. In general, the most credible criticism of existing light-water technology dwells upon the risks (slight but real) of reactor breakdown leading to leakage of radioactive materials, upon sometimes negligent reactor siting in areas of geological instability and, especially, upon the lack of a satisfactory storage program for nuclear wastes. The waste disposal issue is especially disturbing. Despite President Carter's urging that Washington quickly develop plans for the safe disposal of currently generated nuclear wastes, provisions are still so inadequate that a recent review by the California Energy Resources Conservation and Development Commission called federal arrangements "a disaster area." A commission spokesman concluded: "The Government is anticipating a 1,000-reactor economy without really being able to tell you how they're going to deal with the waste."[16]

It is the prospect of the new breeder technology, however, that propels the attack on nuclear technology to a different order of debate. In environmental circles, discussions suggest that the breeder technology will be the leading edge of an onrushing police state. A recent statement on nuclear technology by the Environmental Agenda Task Force sponsored by the Rockefeller Brothers Fund synthesizes the opinion of 12 leading conservationist organizations and captures their doomsday vision:

> Even modestly effective [nuclear plant] safeguards will infringe upon or abrogate traditional civil liberties through surveillance, infiltration, private armies, and extended policy powers. Protecting the guardians of nuclear waste from social unrest, strikes, economic pressures, and wars implies a rigid and hierarchical social structure. An ability to make political decisions about nuclear hazards is so difficult that governments are tempted to bypass the uncertainties of the democratic process in favor of a secretive, elitist technocracy. In short, the technical imperatives of nuclear power are incompatible with the political imperatives of a free society . . .[17]

While addressed specifically to plutonium-based technologies, the indictment is meant to cover *all* variations of nuclear power generation. These concerns, which betray a deep distrust of the political institutions responsible for developing and regulating nuclear technology, are difficult to dispel because they are speculative and impossible to resolve by appeal to technical or scientific data. Unlike objections to existing technology based on specific engineering deficiencies and therefore amenable to settlement through technical innovation, these ideological indictments are beyond the ability of the scientific community to answer. So sweeping and intense a philosophical opposition to nuclear technology strongly implies, as some observers have suspected, that many opponents of existing nuclear generating arrangements truly wish to stop, at once and finally, any continuing expansion of this energy sector even as they assert they are only trying to assure greater "safety" for the public and the environment.

This intense opposition to nuclear power, whatever the justification, forces ecologists to pose for themselves an apparent dilemma illustrating how intimately nuclear and coal policy are associated. If nuclear power generating capacity in the United States does not expand to the magnitude anticipated within the next several decades—a deficit almost inevitable unless opposition to nuclear power today is substantially diminished—the most plausible alternative will be greater reliance upon domestic coal production or severe, and probably unacceptable, forced energy conservation. In light of the environmental risks inherent in expanded coal production, categorical opposition to nuclear development would seem to invite but another kind of potential environmental devastation through coal utilization. Environmental leaders, more alert to this prospective dilemma than many of their constituents and the general public, have sought to leap it by advocating limited, low-technology coal utilization only until softer nonfossil fuel technologies and other technological innovations arrive to ease pressure on coal. Again, the Rockefeller task force declares the gospel: "To buy the time for the transition to [softer energy technologies] the United States will need to build a bridge by using fossil fuels briefly and sparingly."[18] According to this logic ". . . coal can fill the sporadic gaps in our transitional fuel economy without doubling the rate of coal mining. Furthermore, this reliance on coal need last only a few decades, and can remain on a modest scale."

In many respects, assertions that new coal production can be used briefly and sparingly as a bridge to a new energy future without heavy nuclear power production seem more fantasy than fact. Given the many public and private institutions committed to ambitious new coal production, it seems more plausible that any future decrease in anticipated nuclear energy production will give further impetus, and credibility, to those urging still greater coal production to rectify the energy deficit. Falling nuclear energy production coupled with abundant coal would seem, in fact, a resource formula almost irresistible to those seeking a justification for development of coal conversion technologies, massive centralized electric generating systems, and other projects heavily utilizing coal reserves. In this perspective the failure of modest new nuclear power sources to develop within the next few decades would make heavy new coal utilization with its associated hard technologies *more* likely by creating a political climate amenable to   hurried, if not desperate, coal utilization strategies. The more urgent the national need for coal seems, the greater the jeopardy to existing environmental standards that appear to impede coal exploitation and, consequently, the greater the risks of irreversible environmental degradation. In short, the amount of environmental degradation created, or risked, through coal development would seem to be magnified by a strategy of adamant opposition to any future nuclear power development.

The character of future domestic nuclear development will not rest, in any case, wholly upon considerations of environmental impact. Domestic nuclear technologies in the future will also depend upon the Carter administration's attitude toward the export and proliferation of nuclear generating technology. This in turn will affect coal production.

The International Implications

The Carter administration's apprehension about the international proliferation of nuclear materials capable of use in manufacturing nuclear bombs colors its attitude toward domestic nuclear energy development. The president has resolutely opposed further significant federal investment in the Clinch River project and in other activities intended to promote "fast-breeder" reactors and other plutonium-based technologies. This opposition to the breeder reactor is widely viewed within the nuclear power industry as a major impediment to its future growth and a virtual guarantee that the industry will be unable to provide the share of national electric power generation once assumed by planners only a few years ago. It seems unlikely, however, that the president will be able to impose moratorium on the fast-breeder development; he may equally fail in other efforts to inhibit the development of a domestic plutonium-based nuclear technology. Such a failure would clearly rekindle interest in the commercial future of nuclear energy domestically and might lead to a diminishing demand for coal-fired generating plants over the next few decades.

The U.S. breeder reactor program seems assured of new federal funding despite White House opposition. When the Congress in late 1977 voted to fund continued operation of the Clinch River project despite President Carter's threat to veto the appropriation, it was obvious that foes of the breeder lacked predictable political clout. Congressional enthusiasm for breeder technology, despite the alleged dangers of nuclear proliferation, seems to grow from a combination of technological egotism and fear that the nation's security would be threatened if other nations won the "breeder race." Congressmen supporting the Clinch River appropriations were quick to point out that it would "enhance American prestige and power" to have the breeder and would keep the United States in the international breeder ball game. If the United States failed to support the project, the ranking minority member of the House Science and Technology Committee claimed: ". . . we won't be in the club anymore. Ladies and gentlemen, they're going ahead all over the world where they have the capacity to do so, to develop a breeder reactor. We're going to put ourselves out of the ballgame."[19] Besides this, many experts assert that U.S. uranium reserves for conventional reactors will ultimately become limited and new fuel cycles will have to be utilized. Further, technology has been among the most valued U.S. commodities in international trade, one extremely useful in dealing with the oil-producing nations with whom the United States currently runs a large trade deficit because of imported petroleum. Developing new nuclear technologies would offer the nation another continuing technological resource to counterbalance its otherwise unsustainable balance-of-payments deficit. It seems almost assured that such a technology, especially one widely desired by many other nations also seeking alternatives to fossil-fuels for energy generation, will be so irresistibly attractive that domestic support will frustrate President Carter's anti-breeder program. The future economic viability of nuclear generated electric power is likely to depend upon American ability to develop a plutonium-based technology, with capacity for fuel reprocessing. A failure to develop such a technology would almost surely signal a long-term raid on domestic coal reserves in an effort to compensate for nuclear power that never materializes.

WAITING IN THE WINGS: COAL CONVERSION

The relative economic advantages of coal and nuclear power in the next several decades will go far in determining whether the United States initiates a major program to develop coal liquefaction and gasification technologies. The technical capacity to convert coal into liquid fuel (including solvent-refined coal and synthetic crude oil, the most commercially promising forms) and into low-, medium-, and high-BTU gas exists in the United States experimentally. These coal conversion technologies were once widely expected to add further commercial appeal to coal and thereby to strengthen U.S. domestic energy resources. Recently coal conversion has lost much of its former luster. Economic studies indicate that synthetic coal fuels would not be economically competitive with other energy sources over the next several decades. Congress has been reluctant to fund major gasification research; liquefaction research has been only marginally supported.

The economic future of coal conversion could readily change in response to governmental decisions about nuclear fuels and coal. A depressed nuclear power industry would intensify demand for coal and other fuels and perhaps increase their price sufficiently to make synthetic coal fuels economically competitive. A major federal investment in coal conversion research could also stimulate private capital flows into the conversion industry with a resulting development of coal conversion plants across the United States. Generally, as the cost of coal and other fuels increases, the attractions of synthetic fuel grow.

The development of a new coal conversion industry would create major environmental problems for the western United States beyond those already described from coal mining and electric generating plants. Western states have great potential to become the nation's coal conversion belt. When coal conversion seemed more promising, more than 20 liquefaction and gasification plants were planned in western states with large coal reserves; if the liquefaction plants had been constructed they would have produced 1.3 trillion cubic feet of synthetic gas or about 8 percent of domestic natural gas production in 1974.[20] Coal conversion plants require large quantities of cooling water; much of this would have to come from western regions where competition for surface and underground water is already intense. Rapid development of coal conversion installations in the western coal belt would require 142 to 263 thousand acre-feet of water per year by conservative estimates and might grow as high as 329 to 719 thousand acre-feet per year.[21] In addition to competing with requirements for municipal, domestic, and agricultural use, this new demand would also compete with water needed to restore stripped land and, perhaps, to supply coal slurries.

Is there sufficient water to meet coal conversion demand? Western agricultural and cattle groups, fearing the answer is no, allege they will be denied future water resources if a new coal conversion industry arises in their area. It is impossible, given all the present uncertainties about the future coal conversion, to provide even a tentative answer to this question. It seems certain, however, that water adequate to all the competing needs that might arise within a few decades in the West—that is, domestic, commercial, agricultural, mining, reclamation, and slurry demand—will have to come in considerable measure from underground areas. This implies a net loss in the volume of underground water continually available to westerners in the coal belt. Whether the stress upon subsurface western water implies major ecological costs is unclear; the potential for ecological damage is real.

The federal government is not currently prepared to fund coal conversion research lavishly nor yet to abandon it. Hence, it waits in the wings. In his NEP President Carter promised that coal conversion research "would continue at substantial levels" and requested authorization of $527 million in fiscal 1978 for continuing conversion research. It is clear that the administration regards coal conversion as a reserve technology for the present, something to be developed and utilized over the long term but available, if necessary, for rapid exploitation should the situation merit.

SUMMARY: SOME FURTHER IMPLICATIONS OF
INTERNATIONAL POLICY

This chapter has emphasized the sensitivity of domestic coal development to concepts of American strategic interest. U.S. definitions of domestic energy security and acceptable modes of nuclear technology transfer will both affect the pace and breadth of new domestic coal development. Beyond the substantive issues discussed, the analysis also emphasizes the extent to which the interdependent economics of coal and nuclear power are powerfully influenced by governmental policy and other nonmarket institutional factors.

Several conclusions from the analysis are particularly relevant to the environmental movement:

• Environmentalists, together with their allies among other   public interest groups, are virtually the only significant domestic pressure group to challenge significantly the assumption that U.S. energy security requires a rapid, concurrent development of both domestic coal and nuclear energy technologies. Except for constraints on the development of these resources which might arise from market forces, this ecological constituency is the only inhibitant to the energy lobby in influencing domestic energy policy.

• Environmentalists may need to yield in their rigid opposition to nuclear power lest they create conditions causing a new raid on coal resources far greater than that now anticipated by the more optimistic coal development partisans. Ecologists apparently believe they can oppose almost all "nuke" development and still control the environmental ravages of new coal mining that might be stimulated by the future deficit in nuclear power. This hope hangs by an extraordinarily slender thread: the presumption that coal development can be contained until softer technologies arise in the United States. This chapter suggests instead that a significant decrease in anticipated nuclear energy production would be more likely to provoke national support for the hard coal development road.

These conclusions underscore the impossibility of separating nuclear policy from coal policy both domestically and in foreign affairs. They also emphasize the very real dangers inherent in the environmental movement if it adopts the psychology of ideological war in dealing with nuclear power.

NOTES

1. NEP, p. ix.
2. On environmentalist viewpoints, see Gerald O. Barney, ed., *The Unfinished Agenda* (New York: Thomas Y. Crowell, 1977).

3. Ibid., p. 21.

4. *New York Times*, December 14, 1974.

5. *New York Times*, March 13, 1976.

6. *New York Times*, October 12, 1977.

7. Committee for Economic Development, *Achieving Energy Independence* (New York: Committee for Economic Development, 1974), p. 43.

8. Senate, Committee on Interior and Insular Affairs, op. cit., p. 65.

9. OTA, p. 4.

10. *New York Times*, November 16, 1975.

11. *New York Times*, May 24, 1977.

12. Ford Foundation, Nuclear Energy Policy Study Group, *Nuclear Power Issues and Choices* (Cambridge, Mass.: Ballinger Publishing Co., 1977), p. 109.

13. NEP, p. 70.

14. Ibid., p. 72.

15. OTA, p. 56.

16. *New York Times*, July 31, 1977.

17. Barney, op. cit., p. 56.

18. Ibid., p. 62.

19. *New York Times*, September 20, 1977.

20. Comptroller General, *Rocky Mountain Energy Resource Development*, op. cit., p. 18.

21. Northern Great Plains Resources Program, op. cit., p. 73.

5

Alternative Futures: Options and Opportunities in Future Coal Development

The United States should not attempt to extricate itself from its current energy difficulties through a massive, increasing new raid on coal resources. Clearly, the nation will burn more coal, perhaps twice as much as presently, in the next decade; there are persuasive economic and security reasons for encouraging such a short-term climb in coal consumption. But a long-term, open-ended national commitment to increasing coal production involves such grave ecological risk and discourages subsequent energy conservation so much that it should be avoided. Instead, the nation should seek to inhibit coal production to a short-term surge while energy conservation and alternative, more efficient new energy systems become major national policies. If the nation is to err in its coal policy, it should err on the side of excessive restraint.

Fortunately we can control the future. It is possible to constrain new coal demand by attacking the uncritical acceptance of future growth projections, by restraining western coal utilization, by controlling the politicization of coal technology development, and through encouraging the vitality of publics favoring conservative coal policies. Collectively, these strategies can probably dampen political pressure for a new coal boom sufficiently to keep coal utilization policies reasonably conservative. But should these or other conservation efforts fail, the United States could well expect a new coal boom with all its implicit risks. To appreciate the nation's coal policy options, it is important to scrutinize with particular care the implicit or understated assumptions supporting current governmental coal planning. Washington's coal production scenarios, epitomized in the Carter NEP, have too often been accepted without sufficiently probing debate. Further, the language of policy analysis often seems a conspiracy to conceal the abundance of policy options available to the nation. Economic factors relating to coal supply and demand, especially, are often treated as if they are immutable to noneconomic forces when in fact they are so vulnerable to alteration by public policy as to be highly tentative and often quite unpredictable. Moreover, predictions of future energy trends are often extrapolations from the past. "Trends are not destiny," Amory Lovins warns.

Many of the public policies likely to stimulate or restrain future U.S. coal utilization are not overtly economic at all but deal, rather, with the institutional arrangements for making coal policy—that is, with governmental structures and powers. This is not surprising, given the great influence exercised by public agencies on the economy, but it points to the continuing significance of political mechanisms as a central strategy in directing future coal production.

GROWTH POLICY

In the language of energy discussion, energy demand is usually a euphemism for economic growth. Projections of future energy demand carry an implied presumption about future levels of economic activity; indeed, such anticipated energy demand is often an expression of growth rates considered *desirable*. Examining current governmental explorations of coal policy, including the Carter NEP, would lead an unwary reader to presume that future energy demand is largely preordained, a sort of first principle to which coal policy must necessarily be hostage. Consultants working for Washington agencies often impart, however unwittingly, an air of massive inevitability to increased coal demand by basing future coal consumption on current and past energy consumption. Thus, the Stanford Research Institute used extrapolations from current energy utilization to conclude "only powerful governmental, environmental, or socioeconomic forces could retard the growth of Western coal, and the economic loss to the nation of retarding Western coal growth or raising Western coal prices would be large in most cases."[1] In effect, governmental planning of future coal needs is largely captive to assumptions about likely or desirable economic growth.

The NEP did not challenge the idea that future U.S. economic growth—and hence energy demands—should be predicated on past trends; neither do most other congressional proposals for coal development. Washington officials are understandably reluctant to raise a spectre of managed growth with all the political liabilities it entails. The NEP, for instance, assumed an annual growth in the gross national product of 4.2 percent.[2] Current national energy strategies also focus primarily on the short range (that is, demand for energy over the next few decades), a procedure which makes more convenient and plausible the acceptance of current economic growth rates as a basis for decisions on future energy production. In any event, extrapolating future economic growth rates from past and present trends, whatever the justification, virtually stifles debate over the wisdom of such an assumption. In the end, commitments to a given growth policy become components of equations fed into demand models for coal and other energy.

It is quite possible to manipulate energy demand by moderating the stimulants to economic growth, or by applying them selectively in time and place. Also, the rate of energy production in the United States will, to a considerable extent, affect economic growth and consequently the nation can deliberately vary growth rates through energy policy. Thus there is no reason to accept currently assumed levels of future energy production as inevitable, but instead they ought to be considered, for purposes of public discussion, another policy issue.

The implications for coal are very direct. The rate of future coal demand can be moderated through public intervention, as well as private action, in the economic marketplace. The most obvious economic sector involved will be the electric utility industry. In the most literal sense, coal demand is electric power driven. More than 70 percent of the nation's current coal output is consumed by utilities; industrial boilers, by comparison, burn a modest 17 percent of output.[3] Electric utilities have been among the nation's most vigorous growth sectors in terms of power demand and the industry's aggressive marketing of its product. Federal and state governments, which regulate the allocation and pricing of electric power, have added an additional impetus to this growth until recently by regulatory policies that kept electric rates low and encouraged high demand by preferential rates to large power consumers. The FPC regulatory mandate, for example, is explicitly promotional: the FPC is charged with "assuring an abundant supply of electric energy throughout the United States with the greatest possible economy and with regard to the proper utilization and conservation of natural resources. . . ."[4] As interpreted by the FPC, a recent study notes, its duty "might be formulated as holding down prices and keeping service coverage up for the benefit of consumers of gas and electricity."[5] State utility commissions, with jurisdiction over most of the nation's power producers, have been equally deferential to cheap, abundant energy goals in the past. Commonly, state regulatory agencies operate under legislative standards that discourage or prohibit agency constraints on the growth of energy demand. Generally, both the FPC and its counterpart state agencies have in the past believed "they should not consider the impact of their actions on total demand for electricity and whether such demand is desirable."[6]

The structure of electric utility financing has also been a strong incentive to aggressive utility promotion of power consumption. The nation's 3,500 electric utilities have been the biggest U.S. industry in terms of capital investment, accounting for over 12 percent of all private spending for plant and equipment. To finance this large capital demand, the industry is the nation's largest issuer of securities; utility revenues, deeply committed to retiring the industry's large bonded debt, are expected to exceed $30 billion by 1980. Considering the financial structure of the industry, it is understandable that this large bonded indebtedness should incite continual promotion of electric power since it is primarily in this way that the debt must be retired. Demand for electric power, increasing at an annual average of about 7 percent over the last several decades, has customarily been treated by the industry as a given. In recent years, some new restraints have been exercised by utilities in their promotional strategies but the industry is extremely reluctant to embrace any programs that seem to imply growth management in future power demand. Typically, industry spokesmen launch expansion drives on waves of grim warnings. Quite recently, the president of the Edison Electric Institute, an industry organization, took precisely this stance. Cautioning that the nation's power supply outlook might soon be "quite threatening," he observed: "The anticipated continued growth of demand in the coming years strongly emphasizes the necessity to construct new generating plants . . . so they will be ready in time to meet consumer needs." Otherwise, he concluded, "there will be electricity shortages." Another industry spokesman predicted "a hair-curling power crisis by 1985."[7]

Electric utility expansion, and the continuing promotion of electric power consumption that inevitably follows, are usually supported economically and politically by large investment firms. A large portion of the utility industry's capital is derived from banks and investment firms whose utility securities, obtained in return for the monies invested, must be protected. It is common, therefore, to observe at both state and federal levels that the nation's utilities, public and private, often work in close political collaboration with large investment houses to promote national power generating systems and future demand for them.

Federal planners have been reluctant until recently to call for major changes in state and federal regulatory procedures and other public laws, encouraging a constantly climbing electric power consumption. Recently, modest efforts have been proposed in the NEP and elsewhere in the guise of energy conservation policies. President Carter, for example, proposed a number of federally required utility rate reforms, including in particular so-called cost-based rates, which would encourage a very moderate decrease in otherwise expected electric power demand by 1985. At the same time, however, the NEP took an attitude toward electric power consumption that could conceivably stimulate more power generation; it is an approach that appears often in current federal energy planning. From a purely economic viewpoint, it is often logical to encourage the substitution of electric power for oil and gas in industrial, commercial, and residential use—a strategy appealing to planners because the nation has abundant coal. Thus, reasoned the president in the NEP: "To the extent that electricity is substituted for oil and gas, the total amounts of energy used in the country will be somewhat larger due to inherent inefficiency of electric generation and distribution. But conserving scarce oil and natural gas is far more important than saving coal."[8] Such an approach is, in effect, an invitation to raid coal resources in order to support growing power demand as a means of conserving oil and gas— a conservation measure only in the narrow sense that it diminishes petroleum and gas consumption by escalating coal and electric energy as a substitute. Such a policy also seems to work at cross-purposes with professed federal efforts to encourage conservation of all energy resources.

Generally, the NEP betrays the common tendency in governmental energy planning to relegate energy conservation to a low priority in the development of new energy sources. Additionally, conservation is usually conceived in terms of voluntary consumer constraint on energy use and other measures, like taxes and regulatory schemes, that modestly dampen industrial and commercial energy demand below future projected levels without any constraints. It is quite possible to attack the growth of power demand, with its heavy resulting stress on coal reserves, much more directly and effectively. These alternative procedures, though politically more difficult, have been extensively examined and publicized primarily by conservationists, ecologists, and other public interest groups.

The most fundamental need in dealing with national energy demand is the formulation of a target growth rate which adequately reflects a balanced concern for economic expansion and conservation of resources. Federal commissions and other special agencies have been created to examine and recommend national policy on other issues of sweeping importance including pornography, crime and violence, employment, national water resource development, and much else. It would be entirely within this tradition, and particularly appropriate in the context of Washington's current concern with the energy crisis, to initiate a relatively brief but searching appraisal of the growth rate issue immediately. One advantage of such a procedure would be in increasing public visibility and debate of the implications of different growth rates. This would also illuminate the rather direct relationship between assumptions of desirable economic growth and the differing impacts upon energy resources. Also, such an investigation would throw into sharp national relief the inadequacy of many current economic indicators used in governmental planning to reveal the true social costs of economic development. Even if such a federal undertaking produced no satisfactory definition of a target growth rate—perhaps the most likely result, considering the fate of most previous special federal commissions—it might well mobilize greater political pressure to moderate the future energy demand that translates into new coal production.

Other federal measures can also be taken to dampen future electric power demand. Federal court decisions generally suggest that the FPC now has the ability to extend its jurisdiction over most generating plants in the country and consequently national energy policy need no longer be a patchwork of policies evolved through 50 state regulatory agencies. Congress can modify the FPC's regulatory responsibilities to include some, or all, of the following:

• a prohibition on all promotional activities intended, directly or indirectly, to stimulate commercial and residential electric power consumption.

• a prohibition of "block rates," sliding rate schedules, and other pricing policies which allocate preferential rates to large power consumers.

• a statutory requirement that the FPC develop and recommend utility policies intended to foster greater conservation and efficiency of energy use as an equal priority with consumer satisfaction.

Such measures by the FPC, or complementary efforts by state regulatory boards, are no substitute for a more comprehensive congressional formulation of a national electric power policy or, more ideally, a declaration of national growth policy responsive to needs for energy conservation. But such regulatory procedures would themselves be a useful inhibitant to new power demand. These existing capacities to control power demand and to vary it within significant margins suggest that prospective demand for coal need not follow some inevitable and very generous growth curve.

SUSTAINING WESTERN REGIONALISM

The disparity between the geographic dispersion of the nation's coal reserves, its population, and its economic activity creates a formidable political problem for the western coal states. Generally speaking, the nation's population growth and coal demand are concentrated in areas of moderate or nonexistent coal supply. Conversely, the western coal states, with more than half the nation's strippable coal reserves, have neither abundant population nor intensive economic activity. Thus, the states which might experience with particular severity the environmental impact of new coal production are not those needing the coal. This anomaly poses a difficult problem in reconciling national needs with local interests. The difference between the western and nonwestern states in relation to coal supply and demand is suggested in Table 5.1, which contrasts the four states having the most abundant strippable coal reserves in the West to the nation's 10 largest coal-consuming states. The contrast between the western coal states and the remaining United States is further accentuated by recent population data. The fastest growing region of the nation is the South and, particularly, the sunbelt South; this area has virtually no significant coal reserves. Between 1970 and 1976 the population of the Old Confederacy, together with West Virginia and Oklahoma, grew at a combined rate of 9.8 percent compared with a U.S. average of 5.6 percent. Florida, a state wholly without fossil fuel reserves, led the population explosion with a 24 percent growth rate.

This contrast between coal producing and consuming states contributes, along with equally sharp disparities among states in other energy resources, to nurturing an intensified energy regionalism in the country. Spokesmen for the western coal states have become extremely sensitive to the danger that they will be exploited by their energy-hungry neighbors. "If we sit passively by," Governor Thomas Judge of Montana observed in 1975, "we will be a colony for the rest of the United States. The wealth of our states will be sent elsewhere and we will be left with the problems of a boom and bust economy."[9] On the other side, a recent Ford Foundation energy study attempted to take a conciliatory attitude toward those states apprehensive of energy exploitation but seemed instead to warn that the nation will have to take priority over the states if a choice must be made. In choosing what energy technologies to develop, it stated, "We believe it

TABLE 5.1

Coal Supply and Demand: Western and Nonwestern States, 1974

States	Percent of U.S. Strippable Coal	Percent of U.S. Coal Consumption	Percent of U.S. Population
Western states with major strippable coal reserves			
Montana	31.3	.6	.4
New Mexico	1.6	1.2	.4
North Dakota	11.8	1.0	.4
Wyoming	17.5	1.1	.01
Total	62.2	3.9	1.21
Largest coal-consuming states			
Ohio	3.0	11.5	5.3
Pennsylvania	.7	10.5	5.8
Indiana	1.4	7.3	2.4
Illinois	9.0	6.5	5.3
West Virginia	4.0	5.6	1.0
Michigan	—	4.9	4.3
Kentucky	6.0	4.2	1.4
Alabama	.7	4.2	1.9
North Carolina	—	3.5	2.4
Tennessee	—	3.1	1.9
Total	24.8	61.3	31.7

Source: *Statistical Abstract of the United States*, 1974; U.S., Department of the Interior, Bureau of Mines, *Minerals Yearbook, 1974.*

is desirable to allow considerable leeway for local preferences." But: "If it should develop that the cumulative effect of local preferences would endanger a reasonable national mix of coal and nuclear power plants, the case for federal preemption [of nuclear plant licensing and siting procedure] would be stronger."[10] The implicit threat extends beyond nuclear power issues: if local preferences for energy development are deemed incompatible with national "necessity," presumably Washington should intervene to impose national priorities. Certainly, allowing local energy development—coal or otherwise—to be largely determined through national institutions puts the western states at a great political disadvantage. The total congressional delegation from the four major western coal states, for instance, is 15 individuals; this is less than the congressional representation from Ohio, the nation's largest coal consuming state.

Growing awareness of regional energy interest well serves the western states by stimulating greater citizen activity to protect western resources from negligent exploitation. The ability of the western public, many newly mobilized, to affect the course of regional energy development will depend in large measure on federal and state provisions for citizen involvement in energy decisions. Here, institutional arrangements for the expression of local preferences in coal utilization planning are extremely critical. Adequate arrangements mean provisions in federal and state law to inform citizens about governmental procedures affecting local coal utilization, to encourage citizens' participation in these procedures and to provide adequate assistance and opportunity for the expression of citizen views. This participation will sometimes be a constraining influence on coal mining, energy conversion, and power generation within the western coal belt. In this respect, it will be a useful counterforce to the nationally generated pressures for western coal development. It will also enable those who must bear most directly, and perhaps permanently, the costs of coal development to force governmental attention to their interests.

Current federal provisions for public involvement in western coal mine development are generally rudimentary when compared to arrangements for public participation in other new federal environmental programs. This is particularly important in the case of the Strip Mining Control and Reclamation Act which is now the major federal regulatory program affecting western strip mines. The bill explicitly confines opportunities for public involvement largely to public hearings by federal or state agencies prior to making important decisions. Public hearings have often proven inadequate as devices for encouraging governmental attention to citizen opinions. These hearings frequently occur too late in the ripening of important policies to permit citizen ideas to be incorporated; hearings frequently provide insufficient preparation time for groups to obtain and analyze relevant information. Frequently the atmosphere of a public hearing is hostile to open and creative exchange of ideas between public officials and citizens. For these reasons, a number of recent federal laws dealing with air and water pollution, the Coastal Zone Management program, and other environmental matters of special interest to citizen groups contain broader mandates encouraging citizen involvement at diverse and important points in policy development and assisting such citizen participation in a variety of constructive ways.[11] By placing on the affected agencies an affirmative responsibility to promote citizen involvement, rather than placing the burden of initiating activity on citizen groups, these newer laws often encourage agencies to stimulate citizen interest in their programs.

Lacking such a broad participation mandate in the bill itself, other opportunities for citizen participation in administration of the Strip Mining Control and Reclamation Act must depend upon administrative regulations developed by the Department of the Interior and by states participating in the program. Recently the department has revised its administrative regulations to encourage greater receptivity to citizen opinions, broader opportunities for citizen involvement, and somewhat greater resources to assist public participation activities. Nonetheless, the department's long tradition of solicitude for mining interests and its previous indiffernece to environmental groups interested in broadening opportunities for public involvement in departmental activities both suggest that the department is likely to be less than resourceful in voluntarily encouraging more citizen involvement in its procedures.

At the moment, most states likely to assume responsibility for implementing the new federal legislation have provisions, varying enormously in character, for public involvement in their administrative processes. Commonly, these provide for public notice and hearing prior to major administrative decisions and, in most cases, for public appeal from administrative rulings. Relatively few states have written into their administrative procedures a broad and general requirement for public involvement in the administrative process, however, which could serve as a framework for generous public involvement in the implementation of strip mine regulation laws (whether federal or state).

Considering the many deficiencies in current provisions for citizen involvement in coal mine development, a strong case exists for federal and state action to provide greater latitude for this participation. This is one way to assure that broad federal and state commitments to consult local interests in mine development will be given firm and effective operational meaning. A number of specific measures seem appropriate:

• including the new Office of Surface Mining within the federal "Government in the Sunshine Act." This bill, signed by President Ford in late 1976, currently requires about 50 federal boards and commissions to open their meetings to public observation, to keep transcripts of important proceedings, and generally to open heretofore private procedures to public scrutiny. Many of these agencies are regulatory bodies; including the Office of Surface Mining in the bill would be appropriate and consistent with the law's intent.

• amending the Strip Mining Control and Reclamation Act to include a mandate to the Department of the Interior, or to state agencies acting under authority of the law, to promote and assist public involvement in the program. A useful prototype provision can be found in Section 101(e) of the Federal Water Pollution Control Act Amendments (1972).

• providing statutory authorization to federal and state agencies with strip-mine regulatory authority to allocate modest funds to public interest groups for purposes of educating the state publics on the nature of the new regulatory program affecting them. Workshops and other opportunities for citizen education would seem an especially productive use of such funds.

Opening avenues for broad citizen involvement in the regulation of surface coal mining not only encourages activism among existing groups with local viewpoints on mine development but also stimulates and mobilizes citizen interest—a form of politicizing those affected by coal development. Without an alert, mobilized, and effective citizen participation in the regulatory program, it will be far more difficult for the western states to resist the powerful pressures for rapid, broad, and pervasive development of western coal resources.

RESTRAINING THE POLITICIZATION OF TECHNOLOGY DEVELOPMENT

It is prudent for the United States to continue actively exploring in new or modified existing technologies the potential for more economic and efficient coal utilization. Enhancing the economic attraction of coal in competition with other energy sources and facilitating its most effective extraction or combustion with proper environmental safeguards make considerable sense for the U.S. domestic economy and for its international security. Moreover, there is nothing inherently incompatible between the continuing encouragement of coal technology R&D and a conservation approach to future coal utilization. It is possible to improve the nation's coal utilization capacities while still exploring alternative energy technologies. Government can presently encourage more coal utilization without committing the nation to long-term, constantly escalating coal dependency.

It is important, however, that in developing future coal technologies—particularly goal gasification and liquefaction, slurry pipelines, and centralized utility complexes—procedures be utilized that maximize a growing, open-ended national coal dependency. This means especially that technology development should keep options for future national energy use as open as possible, should constantly expose to public scrutiny the implications of new technologies, and should aim primarily at the midterm growth of coal utilization—that is, for the period between roughly 1977 and the year 2 000. Essentially, this is an argument that the United States deliberately avoid any hard-road energy scenarios as described in Chapter 2. The implications of this road have been elaborated. The importance of keeping governmental investment in coal technology development modest in magnitude and restrained in scope needs reiteration.

The logic of this approach is to avoid, insofar as possible, an extensive politicizing of future coal technologies. The federal government should not be permitted to largely underwrite the development of such technologies or to determine their sequence of development. In this respect, U.S. experience with nuclear power development should be considered an object lesson in the inherent problems of governmental development of technology R&D. The lessons are several. First, it is difficult, and sometimes impossible, for Washington to curtail or substantially redesign a major technology system once a large public investment has been made. Proponents of the technology commonly cite the large sunk costs to justify further efforts to redeem the investment, but more importantly a large and politically effective infrastructure of private and public agencies commonly evolves to promote the technology; they will exploit whatever institutional channels—presidential, congressional, or bureaucratic—that will effectively protect this interest. President Carter's attempts to curtail development of the domestic breeder reactor by gradually withdrawing federal support from the Clinch River project illuminate this problem. The president's efforts have been continually frustrated by a large congressional majority allied with a coalition of nuclear and electric utilities, the communities benefiting from the project, and military lobbies. The president's late 1977 veto of congressional appropriations for the project is unlikely to stand. Proponents of the project have defended it on grounds that are as eclectic as national defense, national energy shortages, antirecessionary benefits, national image, and much else. But the larger meaning in the struggle is that they have been able to protect the technology so far, largely through political means.

A second lesson is that federal technology funding often elevates political calculations to far more importance in the selection and dispersion of technologies than the economic or technical merits of such decisions. In its extreme form, this leads Congress, the president, and the bureaucracy to subordinate scientific guidance to peripheral attention. A recent federal decision to locate a new $4.4 billion nuclear enrichment facility, creating perhaps 3,700 new construction jobs by 1988, in Portsmouth, Ohio, rather than Oak Ridge, Tennessee, was unabashedly political. [12] Though it would probably cost the taxpayers several hundred million dollars more to locate the plant in Ohio, the president defended the choice primarily with a campaign pledge made to the people of Ohio and upon the apparently greater need for economic revival in Portsmouth.

A third lesson is that extensive federal involvement in technology R&D easily leads to subsidizing industries or technologies. This is seldom a declared intent at the outset of federal involvement; it is often unanticipated. Still, it happens. Again, the Clinch River project is a useful illustration. The breeder reactor has become increasingly controversial; its technical and economic performance, measured by the standard of expectations once entertained for it, are now debatable. Private investors have become extremely wary of pouring in the large capital required to underwrite any future private development of such a reactor technology until the prototype installation is proven. The federal government consequently has become almost the sole national institution with any potential to develop the breeder reactor; powerful pressures are exerted upon Washington to continue the enterprise. By a curious logic, the risks in the breeder reactor have created every more compelling pressures for Washington to create a viable breeder technology.

Finally, federal R&D investments in energy technology generally work to the advantage of the energy lobby, with its strong preferences for constantly expanding energy production and low priorities for energy conservation. There already exist in Washington, in addition to the private sector energy lobby, several large bureaucracies with strong orientations to expansive energy production. This administrative bloc includes the U.S. Department of the Interior and the FPC. To this must now be added the embryonic Department of Energy. It is instructive that the public debate over the safety and reliability of nuclear power was largely forced by private groups and other interests outside governmental circles.

All these observations underscore the fact that politicizing energy development can readily lead to continual promotion of energy technologies on the basis of considerations very often tenuously related, at best, to the intrinsic economies, technical advantages, or long-range implications of such technologies. Such risks are, of course, inherent to some degree in largely private technology development. Nonetheless, it is still desirable to minimize the impact of forces that encourage such developments. Allocating responsibility for coal technology development largely to the private sector would seem to encourage a greater concern for economic and technical effectiveness and would more strongly inhibit the continuing subsidization of questionable technologies than would public responsibility. Moreover, adequate governmental regulation of the undesirable environmental impacts from these technologies could, when properly applied, confine developments to those whose ecological risks are nationally acceptable.

A number of measures can be taken to minimize many of these undesirable effects of federal R&D for energy development. These measures are particularly practical because they draw upon existing institutional, statutory, and scientific resources and would consequently require little additional modification in federal authority and structures:

• President Carter's guidelines for federal energy R&D, as outlined in his NEP, should be the primary standard for future federal investment. In this respect, particular emphasis should be placed upon minimizing, insofar as possible, federal commitments to proving the ultimate commercial viability of technologies; preventing further major investment in technologies which prove economically or environmentally undesirable after initial development; and requiring that a significant investment balance be maintained between allocations for energy production and energy conservation technologies.

• The congressional OTA should be provided with sufficient resources to permit its continuing examination of the social consequences in all new technologies heavily underwritten by the federal government.

• The CEQ should require federal agencies to file comprehensive environmental impact statements in association with any major support for new energy production technologies.

• Private R&D on energy technology—and especially upon energy-saving technologies—should be stimulated by federal tax and investment credits when

possible, instead of federal subsidization or other direct forms of federal support.

   • A special effort should be made by Washington to encourage energy R&D
activities by smaller and newer companies or other institutions not closely associated with existing energy producers of conglomerates. For this purpose, a
special federal investment fund designed to attract and underwrite proposals
from such institutions should be created.

   None of these measures would discriminate against large corporations and
existing energy producers or consumers in seeking federal assistance for new
energy R&D. Indeed, for very sound reasons it is quite likely that the nation's
existing energy sector will continue to receive the major share of federal investment for future energy R&D. Rather, these measures are an attempt to weaken
the direct and persistent energy R&D involvement of the government that encourages the evolution of wasteful, inefficient, and energy-consuming technologies.

NEEDED: A CONTINUING "QUALITY OF LIFE" PUBLIC

   Constraints on future coal utilization, as on other forms of energy use,
will grow in large measure from values and attitudes which have come to be
identified with a particular American public. In recent years, social commen-
tators have observed the gradual emergence of what has been called the quality
of life public.[13] It is overwhelmingly middle class, well educated, and white; it is
predominantly urban or subruban liberal, and politically active. Though its
boundaries are indistinct and shifting, it shares a broad concern with protecting
and promoting cultural, environmental, and aesthetic values in the political
process. What distinguishes these activists from many other politically involved
Americans, past and present, is the current size of this public and the intensity
with which these values, and the political programs expressing them, are pursued
in political affairs. By most estimates, the Americans deeply committed to
 quality of life issues still constitute a minority of U.S. adults—perhaps no
more than 25 percent of all American adults, by the most generous definition,
could be members of this quality of life constituency.

   The political impact of this significant, and perhaps growing, public is
considerable, however. These Americans are frequently numerous, and often
extremely energetic, within a variety of public interest groups such as the Nader
organizations, Common Cause, or the Sierra Club. They populate in large number the citizen organizations most vigorously exploiting new opportunities for
public involvement in the governmental process. They are, perhaps, the single
most visible and aggressive component within the many groups currently oppos-

ing the future development of nuclear generating plants. In the context of coal policy, this quality of life public is often the backbone of organizations promoting restraint on future coal and other energy development and greater attention to energy conservation; it is a public that responds readily and enthusiastically to future soft energy scenarios for America. Quite often, environmentalists are essentially those Americans whose quality of life concern is expressed with special emphasis in ecological affairs.

We have repeatedly observed that environmentalists and other public interest groups have been among the most vigorous and effective political organizations articulating opinions about massive new energy production within the nation. Indeed, such groups are often the *only* effective counterforce. Beyond the many specific measures suggested in this chapter to preserve and promote attractive alternatives to unlimited future energy production, it is important that this quality of life public remain vital in the nation's political life. Short of a catastrophe, scandal, or other national event of such force that it compels broad public attention to energy conservation, the nation will probably have to depend upon the quality of life public, and its close allies, to create and sustain political pressure for restraint on energy development. In effect, this may well be the public that permits the nation to keep its energy options open over the next few decades.

For these reasons, it is important to encourage public policies that assist public interest groups, environmentalists, and others embodying the quality of life ethic to remain active and visible in national, state, and local political life. Such politics would include generous provision for: first, public participation in the administration of national, state, and local energy programs; second, direct and indirect financial assistance to such groups in their efforts to publicize major national issues (for example, governmental grants to such groups for purposes of creating public information materials, workshops, and media presentations on energy issues); third, continued tax benefits to groups and their financial supporters; and fourth, public agencies which explore and publicize the social consequences of energy policy. It is conceivable that this quality of life public, presently modest in strength, is the precursor of a growing American citizen movement committed to balancing traditional American preoccupations with economic growth and material well-being with attention to moral, ethical, and aesthetic values. At the moment, it is evident only that this public now serves in a very crucial sense as a social conscience forcing Americans to weigh, against their will, the consequences of the growth ethic and its expression in energy policy.

SUMMARY

The thrust of this chapter is that the United States still has numerous options open in developing coal resources. These options will be narrowed, and some entirely foreclosed, if coal development is considered only a hostage to the projected national growth rate and current governmental approaches to coal development are unchallenged. Within the next few years, U.S. coal policy will be in large measure ordained. For all these reasons, the next half-decade will probably be decisive. It will determine whether the undeniable benefits in future coal development will be sufficiently balanced against the grave risks so that the nation can exploit its most abundant energy resource without inflicting upon the next generation a ravaged environment and resources depleted beyond recovery.

NOTES

1. Stanford Research Institute, *A Western Regional Energy Development Study: Economics Vol. 1* (Washington, D.C.: Council on Environmental Quality, 1976), available from National Technical Information Service as Document No. PB-260 835.

2. OTA, p. 192.

3. Consumption figures may be obtained from U.S., Department of the Interior, Bureau of Mines, *Minerals Yearbook*.

4. On the commission's authority, see Stoel, op. cit., pp. 961–62; and Association of the Bar of the City of New York, Special Committee on Electric Power and the Environment, *Electricity and the Environment* (St. Paul, Minn.: West Publishing Co., 1972), pp. 76–92.

5. Breyer and MacAvoy, op. cit., p. 2.

6. Ibid., p. 91.

7. *Gainesville Sun*, January 5, 1978.

8. NEP, p. xii.

9. *New York Times*, September 5, 1975.

10. Ford Foundation, Nuclear Energy Policy Study Group, op. cit., p. 28.

11. Walter A. Rosenbaum, "The Paradoxes of Participation," *Administration and Society* 8 (November 1976): 355–84.

12. *New York Times*, May 26, 1977.

13. This public is discussed in Ronald Inglehart, *The Silent Revolution* (Princeton, N.J.: Princeton University Press, 1977).

Index

105

About the Author

WALTER A. ROSENBAUM is Professor of Political Science at the University of Florida, Gainesville. Formerly a consultant to the Environmental Protection Agency and to the National Science Foundation, he was during 1974–75 a Public Administration Fellow.

Dr. Rosenbaum received his Ph.D. from Princeton University. He is the author of *Political Opinion and Electoral Behavior, Analyzing American Politics, The Politics of Environmental Concern, Political Culture and Public Opinion*, and the forthcoming *Death of Abundance*.

Related Titles
Published by
Praeger Special Studies

ENERGY USE AND CONSERVATION INCENTIVES:
A Study of the Southwestern United States

> William H. Cunningham and
> Sally Cook Lopreato

ALTERNATIVE ENERGY STRATEGIES: Constraints
and Opportunities

> John Hagel III

THE ENERGY CRISIS AND THE ENVIRONMENT:
An International Perspective

> edited by
> Donald R. Kelley

THE ECONOMICS OF NUCLEAR AND COAL POWER

> Saunders Miller

PERSPECTIVES ON U.S. ENERGY POLICY: A Critique of Regulation
American Enterprise Institute Perspectives - III

> edited by
> Edward J. Mitchell

ENVIRONMENTAL REGULATION AND THE ALLOCATION OF COAL:
A Regional Analysis

> Alan M. Schlottmann

*THE ENERGY CRISIS AND U.S. FOREIGN POLICY

> edited by
> Joseph S. Szyliowicz and
> Bard E. O'Neill

*Also available in paperback as a PSS Student Edition.